Recent volumes in MSJ Memoirs

1 **Ivan Cherednik, Peter J. Forrester and Denis Uglov**, Quantum many-body problems and representation theory, 1998 .... JPY4000

2 **Masako Takahashi, Mitsuhiro Okada and Mariangiola Dezani-Ciancaglini (Eds.)**, Theories of types and proofs, 1998 .... JPY5800

3 **Tomotada Ohtsuki**, Combinatorial quantum method in 3-dimensional topology, 1999 .... JPY2000

4 **Atsushi Matsuo and Kiyokazu Nagatomo**, Axioms for a vertex algebra and the locality of quantum fields, 1999 .... JPY2000

5 **Daryl Cooper, Craig D. Hodgson and Steven P. Kerckhoff**, Three-dimensional orbifolds and cone-manifolds, 2000 .... JPY2800

6 **Kong De-xing**, Cauchy problem for quasilinear hyperbolic systems, 2000 .... JPY4000

7 **Vladimir Georgiev**, Semilinear hyperbolic equations, 2000 .... JPY4000

8 **Eric M. Opdam**, Lecture notes on Dunkl operators for real and complex reflection groups, 2000 .... JPY1500

9 **Peter Orlik and Hiroaki Terao**, Arrangements and hyper-geometric integrals, 2001 .... JPY1900

10 **Yuri G. Prokhorov**, Lectures on complements on log surfaces, 2001 .... JPY2500

11 **John R. Stembridge, Jean-Yves Thibon and Marc A. A. van Leeuwen**, Interaction of combinatorics and representation theory, 2001 .... JPY2400

12 **Yves André**, Period mappings and differential equations. From $\mathbb{C}$ to $\mathbb{C}_p$, 2003 .... JPY4500

13 **Shigeaki Koike**, A beginner's guide to the theory of viscosity solutions, 2004 .... JPY2191

14 **Noboru Nakayama**, Zariski-decomposition and Abundance, 2004 .... JPY4286

15 **Valery Alexeev and Viacheslav V. Nikulin**, Del Pezzo and K3 surfaces, 2006 .... JPY2857

16 **Brian H. Bowditch**, A course on geometric group theory, 2006 .... JPY1524

17 **Arkady Berenstein, David Kazhdan, Cédric Lecouvey, Masato Okado, Anne Schilling, Taichiro Takagi and Alexander Veselov**, Combinatorial Aspect of Integrable Systems, 2007 .... JPY2286

18 **Yasutaka Ihara**, On Congruence Monodromy Problems, 2008 .... JPY2667

19 **Joseph Najnudel, Bernard Roynette and Marc Yor**, A Global View of Brownian Penalisations, 2009 .... JPY1733

20 **Danny Calegari**, scl (Stable Commutator Length), 2009 .... JPY3095

21 **Gautami Bhowmik, Kohji Matsumoto and Hirofumi Tsumura (Eds.)**, Algebraic and Analytic Aspects of Zeta Functions and L-functions, 2010 .... JPY2781

22 **Michael Ruzhansky and James Smith**, Dispersive and Strichartz estimates for hyperbolic equations with constant coefficients, 2010 .... JPY2352

23 **Armen Sergeev**, Kähler geometry of loop spaces, 2010 .... JPY3124
24 **Taro Asuke**, Godbillon–Vey class of transversely holomorphic foliations, 2010 .... JPY2114
25 **Hiroshi Sugita**, Monte Carlo method, random number, and pseudorandom number, 2011 .... JPY2333
26 **Shinya Nishibata and Masahiro Suzuki**, Hierarchy of semiconductor equations: relaxation limits with initial layers for large initial data, 2011 .... JPY1952
27 **Suhyoung Choi**, Geometric structures on 2-orbifolds: Exploration of discrete symmetry, 2012 .... JPY2762
28 **Toshio Oshima**, Fractional calculus of Weyl algebra and Fuchsian differential equations, 2012 .... JPY3286
29 **Takeshi Hirai, Akihito Hora and Etsuko Hirai**, Projective representations and spin characters of complex reflection groups $G(m,p,n)$ and $G(m,p,\infty)$, 2013 .... JPY4029
30 **Tatsuo Nishitani**, Cauchy Problem for Noneffectively Hyperbolic Operators, 2013 .... JPY2771
31 **Satoshi Takanobu**, Bohr–Jessen Limit Theorem, Revisited, 2013 ...... JPY3362
32 **Hiroshi Isozaki and Yaroslav Kurylev**, Introduction to spectral theory and inverse problem on asymptotically hyperbolic manifolds, 2014 .... JPY3908
33 **Masaki Maruyama**, Moduli spaces of stable sheaves on schemes, 2016 .... JPY2337
34 **Martin T. Barlow, Tibor Jordán and Andrzej Zuk**, Discrete Geometric Analysis, 2016 .... JPY2864
35 **Osamu Fujino**, Foundations of the minimal model program, 2017 ..... JPY4664
36 **Soichiro Katayama**, Global solutions and the asymptotic behavior for nonlinear wave equations with small initial data, 2018 .... JPY4637
37 **Kazuki Hiroe, Hiroshi Kawakami, Akane Nakamura and Hidetaka Sakai**, 4-dimensional Painlevé-type equations, 2018 .... JPY3046
38 **Alex Casella, Dominic Tate and Stephan Tillmann**, Moduli spaces of real projective structures on surfaces, 2020 .... JPY2309

Moduli spaces of real projective structures on surfaces

Alex Casella, Dominic Tate and Stephan Tillmann

MSJ Memoirs

Mathematical Society of Japan

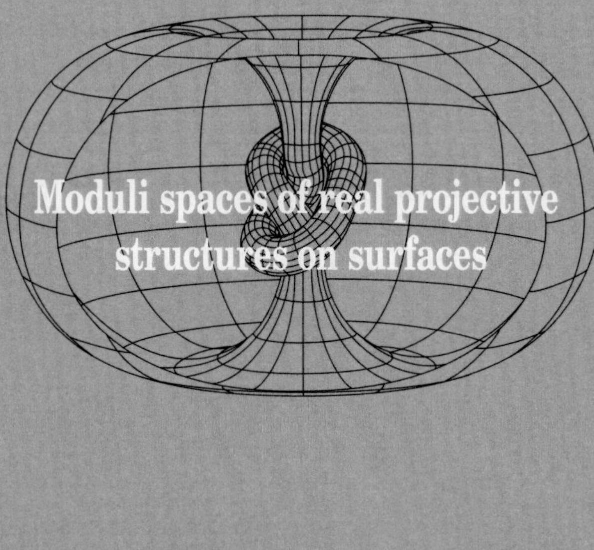

Moduli spaces of real projective structures on surfaces

Alex Casella, Dominic Tate and Stephan Tillmann

Volume 38 2020

Authors
ALEX CASELLA
School of Mathematics and Statistics F07,
The University of Sydney,
NSW 2006 Australia
alex.casella@sydney.edu.au

DOMINIC TATE
School of Mathematics and Statistics F07,
The University of Sydney,
NSW 2006 Australia
dominic.tate@sydney.edu.au

STEPHAN TILLMANN
School of Mathematics and Statistics F07,
The University of Sydney,
NSW 2006 Australia
stephan.tillmann@sydney.edu.au

Keywords: Convex projective surface, hyperbolic surface, moduli space, Teichmüller space, flag, triple ratio.

2010 Mathematics Subject Classifications: Primary 57M50, 51M10, 51A05
Secondary 20F65.

MSJ Memoirs

This monograph series is intended to publish lecture notes, graduate textbooks and long research papers* in pure and applied mathematics. Each volume should be an integrated monograph. Proceedings of conferences or collections of independent papers are not accepted.

Articles for the series can be submitted to one of the editors in the form of hard copy. When the article is accepted, the author(s) is (are) requested to send a camera-ready manuscript.

* long research papers are limited to contributions by MSJ members

Editorial Board

The series *MSJ Memoirs* is published occasionally (3–5 volumes each year) by the Mathematical Society of Japan. The titles and prices of recent volumes are listed on page i of this volume. Each volume of the series can be separately purchased by placing an order with:

- The Mathematical Society of Japan, 1-34-8 Taito, Taito-ku, 110-0016 Tokyo, Japan (from inside Japan)
- World Scientific Publishing, 5 Thoh Tuck Link, Singapore 596224 (from outside Japan)

Preface

The seed for these notes was learning and applying the methods of Fock and Goncharov concerning moduli spaces of real projective structures on surfaces with ideal triangulations. In the process we realised that an expanded treatment would provide an excellent introduction to the study of real projective structures on surfaces and their moduli spaces, and, indeed, a window to the wider fields of real projective manifolds and higher Teichmüller theory.

We have written these notes with two kinds of readers in mind: those who would like to learn about real projective manifolds, and those who have a passing knowledge thereof but are interested in the geometric underpinnings of Fock and Goncharov's parameterisation of moduli space of certain real projective structures. Indeed, we were encouraged to complete this project by both kinds of readers: the former motivated by the simplicity of the constructions, and the latter because many subtle details were not elaborated by Fock and Goncharov and are often misinterpreted.

Much of this material is based on a seminar series at the University of Sydney, and we heartily thank the following people who lectured or participated in the talks: Grace Garden, Montek Gill, Robert Haraway, James Parkinson, and Robert Tang. The last author learned most of what he knows about projective geometry from Daryl Cooper, who introduced him to this rich, puzzling and at times amusing field of research. He also thanks Bill Goldman for first bringing the work of Fock and Goncharov to his attention. We are most grateful to the referee for their careful reading and helpful remarks. We acknowledge support by the University of Sydney, the Commonwealth of Australia and the Australian Research Council.

Alex Casella
Dominic Tate
Stephan Tillmann

Sydney, May 2019

Contents

Introduction

There are different kinds of geometry—they depend on which transformations of space are allowed. Euclidean geometry only allows rigid motions, and this enables us to define distance, angle and volume. In topology, the transformations allowed provide a flexible framework in which the fabric of space is like rubber and which enables the study of the general shape of a space. Our perception of the physical world depends fundamentally on how we imagine the geometry and topology of space-time. This is evidenced by the many physical cosmology theories ranging from Newton, Einstein, and de Sitter's theories to the cyclic models. The objects (or spaces) of geometry and topology are called manifolds, and finding interactions between topological and geometric structures on them holds the key to their study.

To re-imagine the geometry or the topology of a manifold leads to new insights. Some manifolds wear a geometry in one and only one way, while others can wear it in different ways. If one is able to parametrise the different ways in which a geometry is worn, one obtains a *moduli space of geometric structures.*

Given an arbitrary manifold, three fundamental questions are:

- Which geometry should I choose for the manifold?
- How can I parametrise the moduli space of geometric structures?
- What information about the manifold can I gain from the moduli space?

Following Klein's classical viewpoint, the geometry of a manifold is studied via the action of its fundamental group by transformations on a model geometry. This action is called the *holonomy representation*, and deformations of the structure are often studied via the space of conjugacy classes of holonomy representations. One can unroll a manifold in the model geometry just as one can imagine cutting a cylinder and unrolling it in the plane. This leads to a *developing map* for the geometric structure. In the context of these notes, a holonomy representation uniquely determines a developing map for the geometric structure and the so-called Thurston-Ehresmann principle for small deformations applies.

For any 2–dimensional manifold—a surface—classical work of Riemann, Jordan and Möbius from the 19th century gives a clear answer to the first fundamental question, linking the geometry to the topology. A surface has a spherical, Euclidean or hyperbolic structure if its Euler characteristic is positive, zero or negative, respectively. For most surfaces, all closed surfaces for instance, this sufficient condition is also necessary. As a result, hyperbolic geometry (the geometry of constant negative curvature) is the most ubiquitous. The second and third fundamental questions are addressed by Teichmüller theory. Different parametrisations are known for the classical Teichmüller space, which encodes the hyperbolic structures on a surface. For instance, its Fenchel-Nielsen coordinates arise from an explicit construction of the geometric structure, whilst algebraically it is parametrised by certain representations of the fundamental group of the surface into $\mathrm{PSL}(2,\mathbb{R})$.

The algebraic parametrisation of Teichmüller space lends itself to generalisations which describe moduli spaces via spaces of representations into Lie groups. This viewpoint led to the higher Teichmüller theory [14] related to mathematical physics through the work of Witten. The study of moduli spaces connects fields such as Lie theory, dynamical systems, algebraic geometry, representation theory and complex analysis.

The study of Teichmüller space is an active field which enjoys connections with many fields of mathematics, particularly through the omnipresent Riemann surfaces. Classical Teichmüller space has remarkable geometric and analytic structures: a symplectic structure and Kähler structures as discovered by Goldman and Atiyah-Bott, and the dynamics of the action of the mapping class group is well understood. It is a tantalising problem to find analogues of these structure results for other moduli spaces and to use them to gain new information about manifolds or to define new, computable invariants of manifolds. Moreover, classical Teichmüller space has a natural compactification as constructed by Thurston, where points in the compactification are well-understood via degenerations of the geometric structure and measured laminations of the surface. Again, it is a challenge to find compactifications for other moduli spaces that have similar connections to the geometry and topology of the manifold.

These notes are concerned with what might be viewed as a first stepping stone towards higher Teichmüller theory: real projective structures on surfaces. Projective geometry unifies the constant curvature geometries: spherical, Euclidean and hyperbolic geometry. It thus provides a framework for moduli spaces of more general geometric structures in which one can study the passage between different geometries.

The set of all projective structures on a manifold turns out to be too large to yield interesting general results, and these notes therefore focus on *properly convex projective structures* or *strictly convex projective structures*. The foundations of this study were laid by the work of Kuiper (1953, [32, 33]), Vinberg and Kac (1967, [51]) and Benzecri (1960, [2]). Parametrisations, structure and dimension of the moduli spaces were given by Goldman (1990, [21]), Choi-Goldman (1993, [7]) and Marquis (2010, [37]). Finding good parametrisations is the key to applications and further advances of the theory. In these notes, a particular parametrisation due to Fock and Goncharov (2007, [15]) is used to give a self-contained treatment of the key facts about the moduli spaces of these structures on surfaces.

Chapter 1
Overview

Let $\mathbb{RP}^2$ denote the real projective plane and $\mathrm{PGL}(3,\mathbb{R})$ denote the group of projective transformations $\mathbb{RP}^2 \to \mathbb{RP}^2$. A *convex projective surface* is a quotient Ω/Γ, where $\Omega \subset \mathbb{RP}^2$ is an open convex subset and Γ is a discrete (torsion-free) subgroup of $\mathrm{PGL}(3,\mathbb{R})$ that leaves Ω invariant. Examples of convex projective surfaces include Euclidean tori (via the embedding of the Euclidean plane in $\mathbb{RP}^2$ as an affine patch), and hyperbolic surfaces (via the Klein model of the hyperbolic plane). Some convex projective structures on the once-punctured torus are indicated in Figure 1.1.

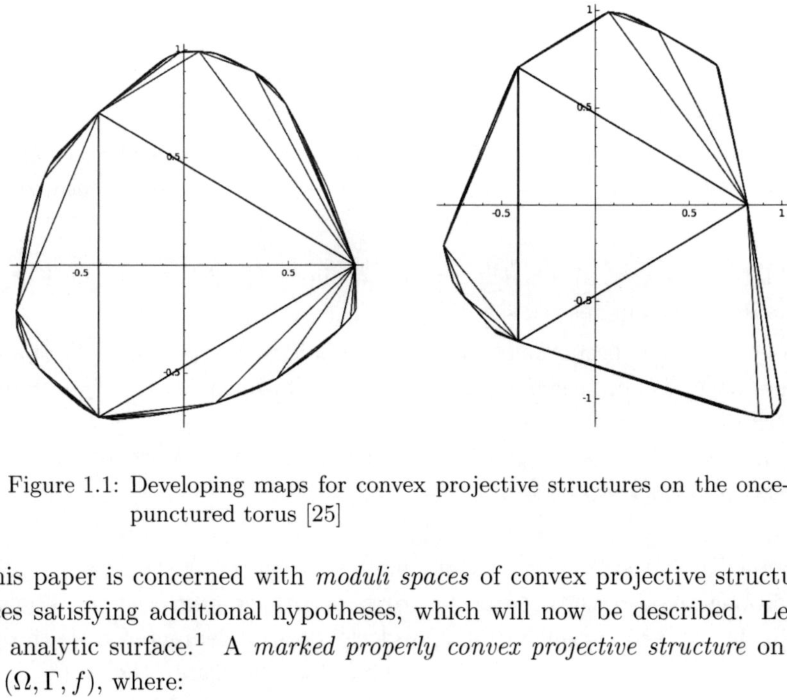

Figure 1.1: Developing maps for convex projective structures on the once-punctured torus [25]

This paper is concerned with *moduli spaces* of convex projective structures on surfaces satisfying additional hypotheses, which will now be described. Let S be a real analytic surface.[1] A *marked properly convex projective structure* on S is a triple (Ω, Γ, f), where:

- $\Omega \subset \mathbb{RP}^2$ is an open subset of the real projective plane which is *proper*, i.e. its closure is contained in an affine patch, and *convex* in that patch;
- Γ is a discrete (torsion-free) subgroup of the group $\mathrm{PGL}(3,\mathbb{R})$ of projective transformations that leaves Ω invariant;
- $f\colon S \to \Omega/\Gamma$ is a real analytic diffeomorphism, where Ω/Γ has a natural real analytic structure, since the action of $\mathrm{PGL}(3,\mathbb{R})$ on $\mathbb{RP}^2$ is real analytic.

[1]Every smooth manifold admits a unique real analytic stiffening [52], and the latter allows us to make use of the concept of a developing map in what is to come.

Fix a universal cover $\widetilde{S} \to S$ and a base point for the fundamental group of S once and for all. The map $f\colon S \to \Omega/\Gamma$ lifts to the *developing map* $\text{dev}\colon \widetilde{S} \to \mathbb{RP}^2$ with image equal to Ω, and induces the *holonomy representation* $\text{hol}\colon \pi_1(S) \to \text{PGL}(3,\mathbb{R})$ with image equal to Γ.

Two marked properly convex projective structures $(\Omega_0, \Gamma_0, f_0)$ and $(\Omega_1, \Gamma_1, f_1)$ are equivalent if there is $T \in \text{PGL}(3,\mathbb{R})$ such that $T\cdot\Omega_0 = \Omega_1$, $T\cdot\Gamma_0\cdot T^{-1} = \Gamma_1$ and f_1 is isotopic to $T^*\cdot f_0$, where T^* is the diffeomorphism $\Omega_0/\Gamma_0 \to \Omega_1/\Gamma_1$ induced by T. The set of all such equivalence classes, which we will refer to as the *moduli space of marked properly convex projective structures*, is denoted $\mathcal{T}_3(S)$. As such, we view a moduli space as a space of objects, and seek suitable parametrisations of this moduli space. The space $\mathcal{T}_3(S)$ has a natural topology, which arises from the weak topology on the set of developing maps $\widetilde{S} \to \mathbb{RP}^2$, viewed as a subspace of $C^\infty_w(\widetilde{S}, \mathbb{RP}^2)$. The subscript in $\mathcal{T}_3(S)$ alludes to the place this moduli space holds in the more general study of discrete and faithful representations of $\pi_1(S)$ into $\text{PGL}(n,\mathbb{R})$; see [14, 34, 23, 3, 22, 30].

In the case of properly convex projective structures on a closed surface, $\mathcal{T}_3(S)$ is homeomorphic with an open subset of the space $\text{Hom}(\pi_1(S), \text{PGL}(3,\mathbb{R}))/\,\text{PGL}(3,\mathbb{R})$ of conjugacy classes of representations of $\pi_1(S)$ into $\text{PGL}(3,\mathbb{R})$. In this case, no further qualification is necessary.

Theorem 1.0.1 (Goldman 1990, [21]). *If S is a closed, orientable surface of negative Euler characteristic $\chi(S)$, then $\mathcal{T}_3(S)$ is diffeomorphic to an open cell of dimension $-8\chi(S)$.*

Goldman [21] gives an explicit set of coordinates for $\mathcal{T}_3(S)$. In these notes, we obtain a different, but related set.

We now fix some useful notation and terminology. Suppose that S_g is a closed, orientable surface of finite genus g. Let $D_1, \ldots, D_n$ be open discs on S_g with pairwise disjoint closures. Choose points $p_k \in D_k$. Then $S_{g,n} = S_g \backslash \{p_1, \ldots, p_n\}$ is a *punctured surface* and we call $\mathcal{E}_k = \overline{D_k}\backslash\{p_k\}$ an *end* of $S_{g,n}$. A *compact core* of $S_{g,n}$ is $S^c_{g,n} = S_g \backslash \cup D_k$. We also write $S_g = S_{g,0}$.

Goldman's theorem states that $\mathcal{T}_3(S_g)$ is an open cell of dimension $16g-16$. The most immediate generalisation of Goldman's theorem to punctured surfaces makes the additional hypothesis that the structures on S have *finite volume.* The volume arises in this context by taking the Hausdorff measure computed from the *Hilbert metric* on the domain Ω, though we remark that there are other ways to define volume that can be used in our context. Denote $\mathcal{T}^f_3(S) \subseteq \mathcal{T}_3(S)$ the subset of all finite volume structures.

Theorem 1.0.2 (Marquis 2010, [37]). *If $S = S_{g,n}$ has negative Euler characteristic and at least one puncture, then $\mathcal{T}^f_3(S)$ is diffeomorphic to an open cell of dimension $16g - 16 + 6n$.*

Note that $S = S_{g,n}$ has negative Euler characteristic if and only if $2g+n > 2$. All ends in a structure of finite volume are cusps in the same manner that non-compact complete hyperbolic surfaces of finite volume have cusps at their ends.

When considering structures of infinite area in $\mathcal{T}_3(S)$, the first consideration is to enumerate the various possibilities for the geometry at each end of S. Consider the interior of a hyperbolic surface F with totally geodesic boundary. One can, in hyperbolic space, equivariantly add a convex collar to the boundary. This collar could be smooth or piecewise linear or a combination thereof. Hence there are infinitely many inequivalent points in $\mathcal{T}_3(F)$ whose structures agree on a compact core of F. In particular, all of these points have the same holonomy.

Returning to the general case of a surface $S = S_{g,n}$, denote by $\mathrm{Hol}(S)$ the set of all holonomies of properly convex projective structures on S. The equivalence relation that defines $\mathcal{T}_3(S)$ descends to the action of $\mathrm{PGL}(3,\mathbb{R})$ by conjugation on $\mathrm{Hol}(S)$, and we denote the quotient by

$$\mathfrak{X}(S) := \mathrm{Hol}(S)/\,\mathrm{PGL}(3,\mathbb{R}).$$

Following Fock and Goncharov [15], we will now *frame* the holonomies. For each end $\mathcal{E}_k$ of S, there is a *peripheral subgroup* $\Gamma_k \leqslant \Gamma$ corresponding to $\mathrm{im}(\pi_1(\mathcal{E}_k) \to \pi_1(S))$. Since $\Gamma_k \cong \mathbb{Z}$, it fixes at least one point in $\mathbb{RP}^2$ and preserves at least one line in $\mathbb{RP}^2$ through that point. A *framing* of the holonomy $\mathrm{hol}\colon \pi_1(S) \to \mathrm{PGL}(3,\mathbb{R})$ consists of the choice, for each end $\mathcal{E}_k$, of such an invariant flag consisting of a point and a line. The set of framed holonomies is denoted

$$\mathrm{Hol}^\times(S_{g,n}) \subset \mathrm{Hom}\left(\pi_1(S_{g,n}), \mathrm{PGL}(3,\mathbb{R})\right) \times \left(\mathbb{RP}^2 \times (\mathbb{RP}^2)^*\right)^n.$$

There is a natural action of $\mathrm{PGL}(3,\mathbb{R})$ on $\mathrm{Hol}^\times(S)$, where the action of an element T of $\mathrm{PGL}(3,\mathbb{R})$ on the dual plane $(\mathbb{RP}^2)^*$ is given by taking the inverse of the transpose of T. We denote the quotient by

$$\mathfrak{X}^\times(S) := \mathrm{Hol}^\times(S)/\,\mathrm{PGL}(3,\mathbb{R}).$$

Theorem 1.0.3 (Fock-Goncharov 2007, [15])**.** *If $S = S_{g,n}$ has negative Euler characteristic and at least one puncture, then $\mathfrak{X}^\times(S)$ is diffeomorphic to an open cell of dimension $-8\chi(S) = 16g - 16 + 8n$.*

The above result is not stated in this form in [15]. What is hidden in the statement is the observation in [15] that there is a section of $\mathrm{Hol}^\times(S) \to \mathfrak{X}^\times(S)$ which consists of so-called *positive representations.* To describe them, we will introduce two constructions of *framed* marked properly convex projective structures on S using developing maps of ideal triangulations. The distinction between these constructions is not made explicit in [15], so we will briefly explain it.

In the above example of the hyperbolic surface F, we concluded that there are infinitely many inequivalent points in $\mathcal{T}_3(F)$ whose structures agree on a compact core of F. This situation was remedied by focussing on the associated holonomies, which were subsequently decorated with flags. Similarly, one can frame a properly convex projective structure as follows. Given (Ω, Γ, f), a framing is the Γ–invariant choice for each peripheral subgroup of a fixed point p in the frontier of Ω and an invariant supporting line to Ω at p. Denote the resulting space $\mathcal{T}_3^\times(S)$. Then there is a natural map $\mathcal{T}_3^\times(S) \to \mathfrak{X}^\times(S)$.

The flags can then be used to construct two different sections for this map. The first section chooses, for each element in $\mathfrak{X}^\times(S)$ the interior of the convex hull of the Γ–orbits of the fixed points in all flags, framed with the corresponding lines. This convex hull will be represented as an *infinite union of ideal triangles,* and the set of these framed structures is denoted $\mathcal{T}_3^+(S) \subset \mathcal{T}_3^\times(S)$. The second section chooses, for each element in $\mathfrak{X}^\times(S)$ the intersection of the Γ–orbits of half-spaces associated to the lines in all flags. This is represented as an *infinite intersection of interiors of ideal triangles* and the set of these with the appropriate framing is denoted $\mathcal{T}_3^-(S) \subset \mathcal{T}_3^\times(S)$. In particular, Theorem 1.0.3 follows from the following result.

Theorem 1.0.4 (Fock-Goncharov 2007, [15])**.** *If $S = S_{g,n}$ has negative Euler characteristic and at least one puncture, then $\mathcal{T}_3^+(S)$ and $\mathcal{T}_3^-(S)$ are diffeomorphic to open cells of dimension $-8\chi(S) = 16g - 16 + 8n$.*

The key lies in a parametrisation of these spaces using an ideal triangulation of the surface S. Fock and Goncharov associate to each triangle in the triangulation a *triple ratio of flags*, and to each oriented edge a *quadruple ratio of flags*, and show how to turn this into homeomorphisms

$$\psi^+ : \mathbb{R}_{>0}^{-8\chi(S)} \to \mathcal{T}_3^+(S) \qquad \text{and} \qquad \psi^- : \mathbb{R}_{>0}^{-8\chi(S)} \to \mathcal{T}_3^-(S),$$

with the property that the following diagram commutes:

$$\begin{array}{ccccc} & & \mathcal{T}_3^+(S) & & \\ & \overset{\psi^+}{\nearrow} & & \overset{\mu^+}{\searrow} & \\ \mathbb{R}_{>0}^{-8\chi(S)} & & & & \mathfrak{X}^\times(S) \\ & \underset{\psi^-}{\searrow} & & \underset{\mu^-}{\nearrow} & \\ & & \mathcal{T}_3^-(S) & & \end{array}$$

For each $\star \in \{+,-\}$, the *monodromy operator* $\mu^\star : \mathcal{T}_3^\star(S) \to \mathfrak{X}^\times(S) \to \mathfrak{X}(S)$ is surjective and generically 6^n–to–1, and corresponds to an action of the direct product of n copies of the symmetric group in three letters. The map $\mu^+ \circ \psi^+ = \mu^- \circ \psi^-$ has a lift $\mathbb{R}_{>0}^{-8\chi(S)} \to \mathrm{Hol}^\times(S)$ that is explicitly described in terms of rational functions where each numerator and denominator is a polynomial with only positive coefficients. These are the so-called *positive representations.*

We remark that the set $\mathcal{T}_3^\pm(S) := \mathcal{T}_3^+(S) \cap \mathcal{T}_3^-(S)$ where the two constructions agree has the property that the composition $\mathcal{T}_3^\pm(S) \to \mathfrak{X}^\times(S) \to \mathfrak{X}(S)$ is surjective and generically 4^n–to–1. There is a more restrictive set $\mathcal{T}_3^{min}(S) \subseteq \mathcal{T}_3^\pm(S)$ of properly convex projective structures, where $\overline{\Omega}$ is *minimal* in the sense that it is the convex hull of the set of all parabolic, attracting and repelling fixed points of peripheral elements. The composition $\mathcal{T}_3^{min}(S) \to \mathfrak{X}^\times(S) \to \mathfrak{X}(S)$ is still surjective and generically 2^n–to–1. Many people we talked to in the initial phase of this project had the impression that Fock and Goncharov's coordinates parametrise the set $\mathcal{T}_3^{min}(S)$. This may arise from the comment after Definition 2.1 in [15] that the set $\mathcal{T}_3^+(S)$ is a 2^n–to–1 cover of the space of non-framed convex real projective structures on S *with geodesic boundary.* In this interpretation, one compactifies those ends of $S = \Omega/\Gamma$ by adding a circle, where a corresponding supporting line in the framing meets the closure of Ω in more than one point.

The duality principle of projective geometry gives rise to natural *duality maps* $\sigma^+ \colon \mathcal{T}_3^+(S) \to \mathcal{T}_3^-(S)$ and $\sigma^- \colon \mathcal{T}_3^-(S) \to \mathcal{T}_3^+(S)$, defined by taking a framed structure to a dually framed dual structure. These are inverse to each other. There also is a natural *dual map* $\sigma \colon \mathfrak{X}^\times(S) \to \mathfrak{X}^\times(S)$ taking the character of a representation to the character of the inverse transpose of that representation, and each flag to the dual flag. We again obtain a commutative diagram:

$$\begin{array}{ccc} \mathcal{T}_3^+(S) & \xrightarrow{\ \mu^+\ } & \mathfrak{X}^\times(S) \\ {\scriptstyle \sigma^+}\big\downarrow\big\uparrow{\scriptstyle \sigma^-} & & {\scriptstyle \sigma}\big\updownarrow \\ \mathcal{T}_3^-(S) & \xrightarrow{\ \mu^-\ } & \mathfrak{X}^\times(S) \end{array}$$

In particular, the action of duality on the coordinate space $\mathbb{R}_{>0}^{-8\chi(S)}$ is given by

$$(\psi^-)^{-1} \circ \sigma^+ \circ \psi^+ = (\psi^+)^{-1} \circ \sigma^- \circ \psi^-.$$

Choosing different triangulations yields transition functions between the different parametrisations of $\mathcal{T}_3^+(S)$ that are again rational functions, where each numerator and denominator is a polynomial with only positive coefficients. Hence $\mathcal{T}_3^+(S)$ inherits an analytic structure from the parametrisation by $\mathbb{R}_{>0}^{-8\chi(S)}$. With respect to this, the monodromy map μ^+ is real analytic and the the set $\mathcal{T}_3^f(S)$ is a real algebraic submanifold of $\mathcal{T}_3^+(S)$. The same remarks apply to $\mathcal{T}_3^-(S)$, and the duality maps σ^+ and σ^- are also real analytic. We show that $\mathcal{T}_3^f(S)$ is a smooth submanifold of $\mathcal{T}_3^-(S)$, and thus has a well-defined analytic structure.

We will give a proof of Theorem 1.0.4 in §4 using the methods of [15]. We also highlight various properties of the moduli space, including an explicit description of the action of the direct product of symmetric groups, the duality map and a *cloverleaf chart* for developing maps. In §5 we describe the action of the mapping class group and characterise the geometry of the ends using the coordinates.

As consequences of Theorem 1.0.4 we then deduce Goldman's Theorem 1.0.1, Marquis' Theorem 1.0.2, as well as a classical result due to Teichmüller about the moduli space of finite area hyperbolic structures on S, subsequently called Teichmüller space, in §6.

As a last application, we focus on the Poisson structure on the moduli space $\mathcal{T}_3^+(S)$ in §6.4. Natural symplectic structures and Poisson structures may often be used to distinguish between diffeomorphic spaces and open the door to the world of integrable systems; see Audin [1] for a discussion in the context of this paper. In the case of closed surfaces, Goldman [17] gives a completely general result via the intersection pairing. Let G be a Lie group preserving a non-degenerate bilinear form on its Lie algebra, and let $\mathfrak{X}_G(S) := \mathrm{Hom}(\pi_1(S), G) /\!\!/ G$ be the *G–character variety* of S. Then Goldman [17] showed that $\mathfrak{X}_G(S_{g,0})$ has a natural symplectic structure. In the case of a punctured surface, where a natural symplectic structure is not available on $\mathfrak{X}_G(S_{g,n})$, it can still be foliated by symplectic leaves making it into a Poisson variety [24]. For $G = \mathrm{SL}(m, \mathbb{R})$, character varieties of surfaces with the same Euler characteristic are isomorphic as varieties, while they may be distinguished using their Poisson structure. A natural question to ask is whether this also applies to the moduli space $\mathcal{T}_3^+(S)$. Following Fock and Goncharov [15], we explicitly construct a Poisson bracket on $\mathcal{T}_3^+(S)$. The monodromy map μ^+ induces an injective homomorphism μ^* from the space of smooth functions on $\mathfrak{X}_3(S) := \mathfrak{X}_{\mathrm{SL}(3,\mathbb{R})}(S)$ to that on $\mathcal{T}_3^+(S)$. When we restrict to the trace algebra $\mathrm{Tr}(\mathfrak{X}_3(S))$, we have the following result (see Theorem 6.4.9 for details):

Theorem 1.0.5. *The induced map μ^* is a Poisson homomorphism between $\mathcal{T}_3^+(S)$ and $\mathfrak{X}_3(S)$.*

We provide a detailed geometric proof of this result, which appears to be known to experts (see, for example, Audin [1] or Nie [41]), but the homomorphism does not appear to have been described explicitly before.

Chapter 2
Projective Structures on Surfaces

According to Felix Klein's influential Erlanger program of 1872,

Geometry is the study of the properties of a space,
which are invariant under a group of transformations.

In the framework of Klein, the familiar Euclidean geometry of the plane consist of the 2–dimensional Euclidean space and its group of isometries. In general, a *geometry* is a pair (G, X), where X is a (sufficiently nice) space and G is a (sufficiently nice) group acting on the space. Geometric properties are precisely those that are preserved by the group. A geometry in Klein's sense may not allow the concepts of distance or angle that are the familiar starting points of Euclidean geometry. Examples of such geometries include affine geometry and projective geometry.

If $H \leqslant G$ is a subgroup, then H corresponds to the group of transformations of a more *rigid* structure on X. This is called a *stiffening* of (G, X), since an (H, X) structure inherits the geometric properties of (G, X), but may have more. Examples of stiffenings are:

$$\text{Euclidean geometry} \xrightarrow{\text{is a stiffening of}} \text{affine geometry} \xrightarrow{\text{is a stiffening of}} \text{projective geometry.}$$

The contents of projective geometry arises from the structure of the underlying vector space. This allows the concepts such as lines, conics and quadrics. Affine geometry allows the additional notions of parallel lines and ratios of distances along parallel lines, and Euclidean geometry allows the additional notions of distance and angle. Figure 2.1 shows a manifestation of stiffening in this context.

A subtle point is that Euclidean and affine transformations leave a subspace of projective space invariant, and this invariant subspace plays a role in the definition of the additional geometric properties. In a similar way, hyperbolic geometry and more general Hilbert geometries are encountered as stiffenings of projective geometry in these notes.

2.1 Projective geometry

The *real projective plane* $\mathbb{RP}^2$ is the space of all lines through the origin in $\mathbb{R}^3$. Scalar multiplication gives an action of the multiplicative group $\mathbb{R}^\times$ on $\mathbb{R}^3\setminus\{0\}$, and we may view $\mathbb{RP}^2$ as the quotient space of this action. A *point* in $\mathbb{RP}^2$ is a line

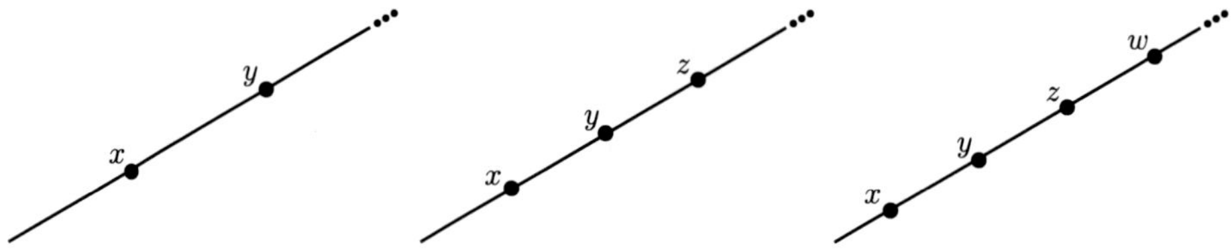

$$\|y-x\| \qquad\qquad \frac{\|z-x\|}{\|y-x\|} \qquad\qquad \frac{\|x-w\|\|z-y\|}{\|x-y\|\|z-w\|}$$

Figure 2.1: Stiffening: *distance* is a meaningful invariant of pairs of points in Euclidean geometry. Pairs of points are equivalent in affine geometry, but there is a *ratio* of three points on a line. Such triples of points are equivalent in projective geometry, but there is a *cross ratio* of quadruples of points on a line.

through the origin in $\mathbb{R}^3$ and it is denoted by a column vector

$$P = \begin{bmatrix} x \\ y \\ z \end{bmatrix} = [x : y : z]^{\mathsf{T}} = \{\lambda(x, y, z)^{\mathsf{T}} \mid \lambda \in \mathbb{R}^{\times}\} \in \mathbb{RP}^2.$$

A *line* in $\mathbb{RP}^2$ is a plane through the origin in $\mathbb{R}^3$ and it is denoted by a row vector

$$\ell = [a : b : c] = \{(x, y, z)^{\mathsf{T}} \in \mathbb{R}^3 \mid ax + by + cz = 0\} \subset \mathbb{RP}^2.$$

The action of $\mathrm{GL}(3, \mathbb{R})$ on $\mathbb{R}^3$ takes lines through the origin to lines through the origin, and descends to a natural action on $\mathbb{RP}^2$. In this way, every linear transformation induces a *projective linear transformation.*

In every dimension, we let $I = \mathrm{diago}(1, \ldots, 1)$ be the identity matrix. The subgroup of non-zero scalar transformations $\mathbb{R}^{\times} \cong \{\ \lambda I \mid \lambda \in \mathbb{R}^{\times}\ \} \lhd \mathrm{GL}(3, \mathbb{R})$ is the kernel of this action on $\mathbb{RP}^2$, and we define the *projective linear group* as the quotient

$$\mathrm{PGL}(3, \mathbb{R}) = \mathrm{GL}(3, \mathbb{R})/\{\ \lambda I \mid \lambda \in \mathbb{R}^{\times}\ \}.$$

We therefore have the exact sequence

$$\{1\} \to \mathbb{R}^{\times} \to \mathrm{GL}(3, \mathbb{R}) \to \mathrm{PGL}(3, \mathbb{R}) \to \{1\}.$$

Going back to Klein's definition of geometry:

Projective geometry is the study of the properties of $\mathbb{RP}^2$,
which are invariant under $\mathrm{PGL}(3, \mathbb{R})$.

The contents of projective geometry thus arises (at first) from the linear algebra of the vector space $\mathbb{R}^3$.

We say that a set of points is *collinear* if there is a line containing all points. Similarly, a set of lines is *concurrent* if the lines have a point in common. Four points (resp. four lines) of $\mathbb{RP}^2$ are in *general position* if no three are collinear (resp. no three are incident). They are often referred as a *projective basis*, because:

Exercise 1. *The action of* $\mathrm{PGL}(3,\mathbb{R})$ *is simply transitive on ordered 4–tuples of points in general position.*

2.1.1 Duality

The *dual vector space* of $\mathbb{R}^3$ is the vector space $\mathrm{Hom}(\mathbb{R}^3,\mathbb{R})$ of all linear transformations $\mathbb{R}^3 \to \mathbb{R}$. As above, scalar multiplication gives an action of $\mathbb{R}^\times$ on $\mathrm{Hom}(\mathbb{R}^3,\mathbb{R})\backslash\{0\}$, and the quotient space is the *dual space* $(\mathbb{RP}^2)^\star$. As suggested by the notation, a line in $\mathbb{RP}^2$ is naturally identified with an element of $(\mathbb{RP}^2)^\star$. In particular, up to projective transformation, there is a canonical dual map between points and lines of $\mathbb{RP}^2$:

$$\perp\colon \mathbb{RP}^2 \to (\mathbb{RP}^2)^*, \qquad \text{where} \qquad P = \begin{bmatrix} x \\ y \\ z \end{bmatrix} \mapsto [x:y:z] = P^\perp.$$

With this interpretation, a point P belongs to a line ℓ if and only if $\ell(P) = 0$. To avoid introducing more cumbersome notation, we denote by the same symbol also the inverse map $\perp\colon (\mathbb{RP}^2)^* \to \mathbb{RP}^2$.

If $\mathcal{B} = \{B_1, B_2, B_3\}$ is a basis of $\mathbb{R}^3$, then there is a *dual basis* $\mathcal{B}^* = \{f_1, f_2, f_3\}$ of $\mathrm{Hom}(\mathbb{R}^3,\mathbb{R})$ characterised by $f_j(B_k) = \delta_{jk}$ for all $1 \leqslant j,k \leqslant 3$.

An element $T \in \mathrm{GL}(\mathbb{R}^3)$ acts on $(\mathbb{RP}^2)^*$ by:

$$T^*(f) := f \circ T^{-1}.$$

In particular, the matrices with respect to the bases are related by:

$$[T^*]_{\mathcal{B}^*} = ([T^{-1}]_{\mathcal{B}})^\intercal.$$

It follows by duality and Exercise 1 that the action of $\mathrm{PGL}(3,\mathbb{R})$ is simply transitive on ordered 4–tuples of lines in general position.

Exercise 2. *Why is the action defined by* $T^*(f) = f \circ T^{-1}$ *and not (as found in most linear algebra texts) by* $T^*(f) = f \circ T$*?*

2.1.2 Projective coordinates and the projective line

Choosing a basis $\{B_1, B_2, B_3\}$ for $\mathbb{R}^3$ gives *projective coordinates*

$$[t_1B_1 + t_2B_2 + t_3B_3] = [t_1 : t_2 : t_3]^\intercal = [\lambda t_1 : \lambda t_2 : \lambda t_3]^\intercal,$$

where $\lambda \in \mathbb{R}^\times$. Given a line $\ell \subset \mathbb{RP}^2$, one can *choose* projective coordinates so that

$$\ell = \{[x_1 : x_2 : 0]^\intercal \in \mathbb{RP}^2 \mid x_k \in \mathbb{R}, \text{ not all } x_k \text{ equal to } 0\}.$$

An *affine patch* is the complement of a line. The complement of the above ℓ is given by:

$$\mathcal{A} = \mathbb{RP}^2 \setminus \ell = \{[x_1 : x_2 : 1]^\intercal \mid x_k \in \mathbb{R}\}.$$

Dehomogenising identifies the affine patch $\mathcal{A} = \mathbb{RP}^2 \setminus \ell$ with $\mathbb{R}^2$ via its *standard coordinates*:

$$[x_1 : x_2 : x_3]^\intercal \mapsto \left(\frac{x_1}{x_3}, \frac{x_2}{x_3}\right)^\intercal = (y_1, y_2)^\intercal.$$

where the inverse map $\mathbb{R}^2 \to \mathcal{A}$ is given by $(y_1, y_2)^\intercal \mapsto [y_1 : y_2 : 1]^\intercal$. We use the notation $\mathcal{A} = \mathbb{R}^2$ to indicate that some standard coordinates have been chosen for the affine patch $\mathcal{A}$.

The definition of the projective plane generalises naturally to all dimensions, giving the n–dimensional *real projective space* $\mathbb{RP}^n = (\mathbb{R}^{n+1}\setminus\{0\})/\mathbb{R}^\times$ as the space of all lines in $\mathbb{R}^{n+1}$. Here, one again has the induced action of $\mathrm{PGL}(n+1, \mathbb{R})$ on $\mathbb{RP}^n$. In these notes, we are interested in the projective plane $\mathbb{RP}^2$ and projective lines contained in $\mathbb{RP}^2$. If we refer to *the* projective line as a separate entity, it is understood that we refer to $\mathbb{RP}^1$.

We often identify a projective line in $\mathbb{RP}^2$ with $\mathbb{RP}^1$, which (via its projective coordinates) is identified with $\mathbb{R} \cup \infty$ using the map $[x : y]^\intercal \mapsto \frac{x}{y}$.

In what follows we describe a number of stiffenings of projective geometry, where we will use the following definition. Given the subsets $S_1, \ldots, S_m \subset \mathbb{RP}^2$, the following defines a subgroup of $\mathrm{PGL}(3, \mathbb{R})$:

$$\mathrm{PGL}(S_1, \ldots, S_m) := \{T \in \mathrm{PGL}(3, \mathbb{R}) \mid T(S_k) \subseteq S_k \text{ for } 1 \leqslant k \leqslant m\}.$$

Exercise 3. *Let* $\ell \subset \mathbb{RP}^2$ *be a line. Describe* $\mathrm{PGL}(\ell) \subset \mathrm{PGL}(3, \mathbb{R})$.

2.1.3 Affine geometry

The identification of an affine patch $\mathbb{RP}^2 \setminus \ell$ with $\mathbb{R}^2$ also gives an identification

$$\mathrm{PGL}(\mathbb{RP}^2 \setminus \ell) \leftrightarrow \mathrm{Aff}(2) := \{x \mapsto Ax + b \mid A \in \mathrm{GL}(2, \mathbb{R}), b \in \mathbb{R}^2\} \cong \mathrm{GL}(2) \ltimes \mathbb{R}^2,$$

where the embedding of $\mathrm{Aff}(2)$ into $\mathrm{PGL}(3, \mathbb{R})$ is given via the following map to $\mathrm{GL}(3, \mathbb{R})$:

$$(x \mapsto Ax + b) \mapsto \begin{pmatrix} A & b \\ 0 & 1 \end{pmatrix}.$$

It follows that $(\mathbb{RP}^2 \setminus \ell, \mathrm{PGL}(\mathbb{RP}^2 \setminus \ell))$ is identified with the 2–dimensional *affine space* $(\mathbb{R}^2, \mathrm{Aff}(2))$, and hence affine geometry is a stiffening of projective geometry.

In affine geometry, there is no interesting function on pairs of points that is invariant under the action of $\mathrm{Aff}(2)$, but there is an interesting such function on triples of points. This will be shown in the following exercise.

Exercise 4. *Let* l_0 *and* l_1 *be (affine) lines in* $\mathbb{R}^2$ *and* $x, y \in l_0$ *and* $a, b \in l_1$ *be pairs of distinct points. Show that there is* $T \in \mathrm{Aff}(2)$ *such that* $Tx = a$ *and* $Ty = b$.

Moreover, show that any two such affine maps agree on l_0.

Hence deduce that any function $f\colon \mathbb{R}^2 \times \mathbb{R}^2 \to \mathbb{R}$ *which is invariant under* Aff(2) *is constant.*

Next, identify $\mathbb{R}$ *with the affine line* $l_1 = \{(x, 0, 0) \mid x \in \mathbb{R}\}$, *let* l_0 *be any line and* $x, y, z \in l_0$ *be pairwise distinct points. Define* $[x, y, z] = Tz$, *where* $T\colon \ell_0 \to \ell_1$ *is an affine transformation mapping* $Tx = 0$ *and* $Ty = 1$. *Show that if* $l_0 = \mathbb{R}$, *then*

$$[x, y, z] = \frac{z - x}{y - x}.$$

This last line can be paraphrased as: affine geometry allows the concept of "ratios of distances along parallel lines." Also note that the *ratio* can be extended to triples of points of which at least two are distinct.

We recall some familiar affine transformations. Suppose $T = \begin{pmatrix} A & b \\ 0 & 1 \end{pmatrix}$ is given as above.

1. If $A = \begin{pmatrix} 1 & 0 \\ 0 & 1 \end{pmatrix}$ is the identity matrix, then T is a translation.
2. If $A = \begin{pmatrix} 1 & 1 \\ 0 & 1 \end{pmatrix}$ and $b = \begin{pmatrix} 0 \\ 0 \end{pmatrix}$, then T is a horizontal shear by 1.
3. If $A = \begin{pmatrix} \cos\theta & -\sin\theta \\ \sin\theta & \cos\theta \end{pmatrix}$ is a rotation matrix and $b = \begin{pmatrix} 0 \\ 0 \end{pmatrix}$, then T is a counterclockwise rotation by θ about the origin.
4. If $A = \begin{pmatrix} \lambda_1 & 0 \\ 0 & \lambda_2 \end{pmatrix}$ is a diagonal matrix and $b = \begin{pmatrix} 0 \\ 0 \end{pmatrix}$, then T is a scaling. It is a *uniform scaling* or *dilation* if $\lambda_1 = \lambda_2$ and a *non-uniform scaling* otherwise.

Using the above transformations and the real Jordan normal form of elements of $\mathrm{GL}(2, \mathbb{R})$, it easily follows that any affine transformation is a composition of rotations, translations, scalings and shears. We remark that there are other projective transformations that preserve an affine patch but are not affine transformations.

2.1.4 Euclidean geometry

Affine geometry can be further stiffened to *Euclidean geometry.* This again has underlying space $\mathbb{R}^2$, but with the smaller group of transformations

$$\mathrm{Euc}(2) := \{x \mapsto Ax + b \mid A \in \mathrm{O}(2), b \in \mathbb{R}^2\} \cong \mathrm{O}(2) \ltimes \mathbb{R}^2,$$

where $\mathrm{O}(2) = \{A \in \mathrm{GL}(2, \mathbb{R}) \mid A^t A = I\}$ is the orthogonal group.

The *Euclidean metric* on $\mathbb{R}^2$ is given via

$$d_{euc}(x, y) = \|x - y\| = \sqrt{(x - y) \cdot (x - y)},$$

and for each $T \in \mathrm{Euc}(2)$ and all $x, y \in \mathbb{R}^2$, one has $d_{euc}(Tx, Ty) = d_{euc}(x, y)$.

Denote the 2–dimensional Euclidean space $\mathbb{E}^2 = (\mathbb{R}^2, d_{euc})$. Here is a sketch of the fact that $\operatorname{Isom}\mathbb{E}^2 = \operatorname{Euc}(2)$, and that the latter has the claimed structure. To begin with, one shows that every isometry $\alpha \in \operatorname{Isom}\mathbb{E}^2$ can be written in the form $\alpha(x) = Ax + b$, where $A \in \mathrm{O}(2)$ and $b \in \mathbb{R}^2$. To do this, one uses linear algebra to first show that an isometry fixing the origin must be of the form $\alpha(x) = Ax$, where $A \in \mathrm{O}(2)$. A general isometry will take the origin to some $b \in \mathbb{R}^2$, so composing this with the translation by $-b$ gives an isometry fixing the origin.

This decomposition of α into a rotational part and a translational part now leads to the group structure. Denote the translation $x \mapsto x + b$ by T_b. The subgroup of translations is normal in $\operatorname{Isom}\mathbb{E}^2$ and isomorphic to $\mathbb{R}^2$. The isometries fixing the origin also form a subgroup, and this is isomorphic with $\mathrm{O}(2)$. Now every isometry in the above form is written uniquely as an isometry fixing the origin followed by a translation (it can also be written uniquely as a translation followed by an isometry fixing the origin, but this is irrelevant). With the representation $\rho\colon \mathrm{O}(2) \to \operatorname{Aut}(\mathbb{R}^2)$ taking $A \in \mathrm{O}(2)$ to the automorphism $T_b \to AT_bA^{-1} = T_{Ab}$, one has

$$\operatorname{Isom}\mathbb{E}^2 \cong \mathbb{R}^2 \rtimes_\rho \mathrm{O}(2).$$

2.1.5 Cross ratio

Exercise 5. *There exists a unique projective map of the projective line $\mathbb{RP}^1$ to itself taking any three pairwise distinct points on the line to any other three pairwise distinct points.*

As a result of the above exercise, there can be no non-trivial function on pairs or triples of points on $\mathbb{RP}^1$ that is invariant under all projective transformations of $\mathbb{RP}^1$. However, one can define a *cross ratio* of four points as follows.

Let $\ell \subset \mathbb{RP}^2$ be a line. Given $w, x, y, z \in \ell$, with x, y, z pairwise distinct, denote $T\colon \ell \to \mathbb{RP}^1$ the unique projective map such that $Tx = 0$, $Ty = 1$ and $Tz = \infty$. Then define

$$[x, y, w, z] = Tw.$$

Exercise 6. *If $\ell = \mathbb{RP}^1$, this is given by the formula:*

$$[x, y, w, z] = \frac{(x-w)(z-y)}{(x-y)(z-w)}.$$

The cross ratio can be extended to quadruples of collinear points in $\mathbb{RP}^2$ of which at least three are pairwise distinct.

Exercise 7. *Let σ be a permutation on four symbols. Show that there exists a linear fractional transformation Φ_σ such that*

$$[x_{\sigma(1)}, x_{\sigma(3)}, x_{\sigma(3)}, x_{\sigma(4)}] = \Phi_\sigma([x_1, x_2, x_3, x_4]).$$

In particular, determine which permutations leave cross ratio invariant.

A pair (y, w) is *harmonic* with respect to (x, z), in which case (x, y, w, z) is a *harmonic quadruple*, if and only if

$$[x, y, w, z] = -1.$$

Exercise 8. *Let $f\colon \mathbb{RP}^1 \to \mathbb{RP}^1$ be a map. Show that the following are equivalent:*

1. *f is a projective map;*
2. *f is a bijection and takes harmonic quadruples to harmonic quadruples;*
3. *f preserves cross ratio; i.e.*

$$[f(x), f(y), f(w), f(z)] = [x, y, w, z].$$

for all quadruples (w, x, y, z).

Exercise 9. *Let $\mathbb{R}^2$ be an affine patch of $\mathbb{RP}^2$, imbued with the Euclidean metric. Suppose w, x, y, z lie on an affine line such that, in $\mathbb{R}^2$, x and z are the endpoints of a segment on this line containing y and w, and such that y lies between x and w on the line segment. Show that*

$$[x, y, w, z] = \frac{\|x - w\| \cdot \|z - y\|}{\|z - w\| \cdot \|x - y\|}.$$

Notice that the cross ratio in this set-up satisfies $[x, y, w, z] \geqslant 1$.

Exercise 10. *Given $[x], [z] \in \mathbb{RP}^2$, let $[y] = [\lambda x + \mu z]$ and $[w] = [\lambda' x + \mu' z]$. Show that*

$$[x, y, w, z] = \frac{\lambda\mu'}{\lambda'\mu}.$$

Exercise 11. *Show that there is a well-defined cross ratio of four lines in $\mathbb{RP}^2$ incident at a common point.*

2.1.6 Elliptic space

Spherical geometry is usually described on the round 2–dimensional sphere, but the treatment becomes more succinct when one identifies antipodal points; for instance two points on the sphere lie on a unique (spherical) line if and only if they are not antipodal.

The 2–sphere $\mathbb{S}^2$ has a natural Riemannian geometry of constant curvature $+1$, which coincides with the path metric that it inherits as the level set

$$x_1^2 + x_2^2 + x_3^2 = 1$$

in $\mathbb{R}^3$. The antipodal map is an isometry of this metric, and hence the quotient metric space, called *elliptic space*, has a Riemannian geometry of constant curvature $+1$. It follows from its description that elliptic space (with its real analytic structure) is naturally identified with $\mathbb{RP}^2$.

It is well known that no coordinate system on the projective plane is regular everywhere. The line element for the elliptic metric on an affine patch is given by

$$ds^2 = \frac{dy_1^2 + dy_2^2 + (y_1 dy_2 - y_2 dy_1)^2}{(1 + y_1^2 + y_2^2)^2}$$

with respect to the standard coordinates described in §2.1.2 (cf. [5, §38].

The n–dimensional orthogonal group is defined by

$$\mathrm{O}(n) = \{A \in \mathrm{GL}(n, \mathbb{R}) \mid A^t A = I\}.$$

The group of isometries of elliptic space is naturally identified with the projective orthogonal group PO(3), and it follows that

$$\text{elliptic geometry } (\mathbb{RP}^2, \mathrm{PO}(3)) \xrightarrow{\text{is a stiffening of}} \text{projective geometry } (\mathbb{RP}^2, \mathrm{PGL}(3, \mathbb{R})).$$

2.1.7 Topologies and the special linear group

We present three different viewpoints on the topology and differentiable structure of $\mathbb{RP}^2$. We have described $\mathbb{RP}^2$ as the quotient topological space of $\mathbb{R}^3 \backslash \{0\}$ (with the usual Euclidean topology) by the equivalence relation $x \sim y$ if and only if $x = \lambda y$ for some $\lambda \in \mathbb{R} \backslash \{0\}$. The quotient topology from this description coincides with the metric topology that arises from viewing $\mathbb{RP}^2$ as elliptic space, since the metric on $\mathbb{S}^2$ is the path metric from the Euclidean metric. There is a natural analytic atlas with three charts consisting of the following affine patches:

$$\begin{aligned} \mathcal{A}_1 \ni [1 : x_1 : x_2]^\mathsf{T} &\leftrightarrow (x_1, x_2)^\mathsf{T} \in \mathbb{R}^2, \\ \mathcal{A}_2 \ni [y_1 : 1 : y_2]^\mathsf{T} &\leftrightarrow (y_1, y_2)^\mathsf{T} \in \mathbb{R}^2, \\ \mathcal{A}_3 \ni [z_1 : z_2 : 1]^\mathsf{T} &\leftrightarrow (z_1, z_2)^\mathsf{T} \in \mathbb{R}^2. \end{aligned}$$

The transition functions are given by rational functions, and hence analytic.

The elements of $\mathrm{GL}(3, \mathbb{R})$ are real analytic maps $\mathbb{R}^3 \to \mathbb{R}^3$, and descend to analytic maps $\mathbb{RP}^2 \to \mathbb{RP}^2$. Last, the group $\mathrm{GL}(3, \mathbb{R})$ has a natural subspace topology from $\mathbb{R}^9$, which makes it into a Lie group, and $\mathrm{PGL}(3, \mathbb{R})$ is given the quotient topology.

Exercise 12. *The described quotient topology on* $\mathrm{PGL}(3, \mathbb{R})$ *is the same as the compact-open topology on* $\mathrm{PGL}(3, \mathbb{R})$.

Exercise 13. *Fix a projective basis* $\mathcal{B}_0$ *of* $\mathbb{RP}^2$*, and define*

$$h\colon\ \mathrm{PGL}(3, \mathbb{R}) \to \{\mathcal{B} \mid \mathcal{B} \text{ is a projective basis of } \mathbb{RP}^2\} \subset (\mathbb{RP}^2)^4$$

by $h(\varphi) = \varphi(\mathcal{B}_0)$. *Then* h *is a homeomorphism (where* $(\mathbb{RP}^2)^4$ *is given the product topology).*

There is a natural analytic isomorphism $\mathrm{PGL}(3,\mathbb{R}) \to \mathrm{SL}(3,\mathbb{R})$ defined by

$$A \mapsto (\det A)^{-1/3}\, A.$$

We will henceforth work with the group $\mathrm{SL}(3,\mathbb{R})$, and note that it is isomorphic as an analytic group with $\mathrm{PGL}(3,\mathbb{R})$. In particular, any arguments involving continuity, topology or analyticity that arise in these notes are not affected by working with the special linear group. Moreover, this not only simplifies notation, but also allows us to make use of the concepts of *trace* and *eigenvalues.*

2.2 Projective structures

Before we come to a rigorous definition of a geometric structure on a surface, using developing map and holonomy, we will motivate this by some examples.

2.2.1 Euclidean structures on the torus

The *standard Euclidean torus* arises as the quotient $\mathbb{E}^2/\mathbb{Z}^2$, where $\mathbb{Z}^2 \leqslant \mathbb{R}^2$ acts by translations. With the identification of $\mathbb{R}^2$ as an affine patch in $\mathbb{RP}^2$, the standard Euclidean torus corresponds to the action $\mathbb{Z}^2 \curvearrowright \mathbb{R}^2 \subset \mathbb{RP}^2$ given by the monorphism

$$\mathbb{Z}^2 \to \mathrm{SL}(3,\mathbb{R}) \qquad \text{defined by} \qquad \begin{pmatrix} n \\ m \end{pmatrix} \mapsto \begin{pmatrix} 1 & 0 & n \\ 0 & 1 & m \\ 0 & 0 & 1 \end{pmatrix}.$$

More generally, if Γ is a subgroup of translations of $\operatorname{Isom}\mathbb{E}^2$ generated by two linearly independent translations, then $\mathbb{E}^2/\Gamma$ gives a Euclidean structure on the torus. Such a subgroup Γ is called a *lattice* in $\operatorname{Isom}\mathbb{E}^2$. Up to rotation and dilation, we may assume that Γ is generated by translations

$$T_1 = \begin{pmatrix} 1 & 0 & 1 \\ 0 & 1 & 0 \\ 0 & 0 & 1 \end{pmatrix} \qquad \text{and} \qquad T_2 = \begin{pmatrix} 1 & 0 & x \\ 0 & 1 & y \\ 0 & 0 & 1 \end{pmatrix},$$

where x and y are positive. For convenience, we identify the affine patch with the complex plane $\mathbb{C}$. Then the translations generating Γ are $\rho_1(z) = z + 1$ and $\rho_2(z) = z + \omega$, where $\omega = x + iy \in \mathbb{H}^2 = \{w \in \mathbb{C} \mid \mathrm{Im}(w) > 0\}$. Implicitly, this choice of ordered generators for the lattice corresponds to a preferred choice of oriented basis of the fundamental group of the torus. This is referred to as a *marking.* Fixing $\pi_1(S^1 \times S^1, p) = \langle \alpha, \beta \rangle$, we have the holonomy representation $\pi_1(S^1 \times S^1, p) \to \operatorname{Isom}\mathbb{E}^2$ generated by $\alpha \mapsto T_1$ and $\beta \mapsto T_2$, and the point $p \in S^1 \times S^1$ corresponds to the image of the origin in $\mathbb{C}/\Gamma$. Moreover, the marking determines a map $S^1 \times S^1 \to \mathbb{C}/\Gamma$ up to isotopy fixing p. This is illustrated in Figure 2.2.

The up-shot of this discussion is that the space of marked Euclidean structures (up to rotation and dilation) on the oriented torus is parametrised (up to isotopy) by

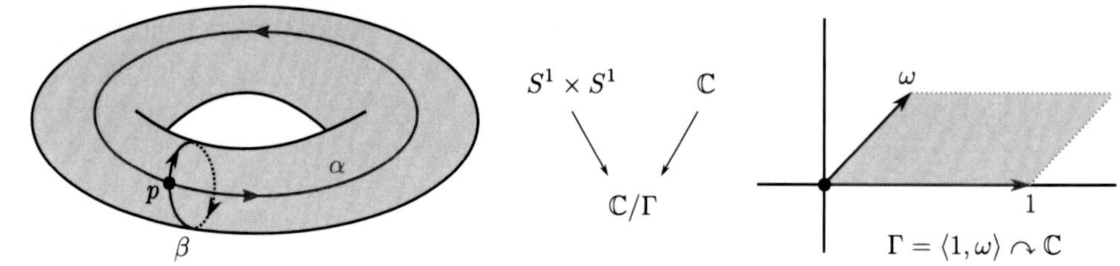

Figure 2.2: Marked Euclidean structure on the torus

$\omega \in \mathbb{H}^2$. To paraphrase this: $\mathbb{H}^2$ is the moduli space of marked Euclidean structures (up to rotation and dilation) on the oriented torus. The *topology* of $\mathbb{H}^2$ allows us to study continuous families of Euclidean structures on the torus, and we see that geometric properties, such as the translation length of a given group element, vary continuously in such a family. Similar observations apply to the smooth or real analytic structure of $\mathbb{H}^2$.

A moduli space of marked structures is often termed a *Teichmüller space.* Any given lattice has infinitely many different bases, and these give rise to different points in Teichmüller space. In the case of the torus, the equivalence relation is not difficult to determine. Given any two *oriented* bases of $\mathbb{Z}^2$, there is an element

$$A = \begin{pmatrix} a & b \\ c & d \end{pmatrix} \in \mathrm{SL}(2, \mathbb{Z})$$

taking one to the other. Now two oriented lattices $\Gamma_1 = \langle 1, \omega_1 \rangle$ and $\Gamma_2 = \langle 1, \omega_2 \rangle$ are equal if and only if there is $A \in \mathrm{SL}(2, \mathbb{Z})$ as above such that

$$\omega_2 = \frac{a\omega_1 + b}{c\omega_1 + d}.$$

So the moduli space of unmarked Euclidean structures on the torus naturally corresponds to the space $\mathbb{H}^2 / \mathrm{SL}(2, \mathbb{Z})$. A fundamental domain for this in $\mathbb{H}^2$ is shown in Figure 2.3, and the quotient orbifold is shown in Figure 2.4.

2.2.2 Developing map and holonomy

Let S be a real analytic surface. A *real projective atlas* of S is a collection of maps

$$\{\Psi \colon U \to V \subseteq \mathbb{RP}^2\},$$

where the sets U are open and their union covers S, and the maps Ψ are real analytic diffeomorphisms onto open sets V. The space of atlases may be naturally ordered by inclusion. A *real projective structure* on S is a maximal real projective atlas with the property that all transition maps

$$\Psi_j \circ \Psi_i^{-1} \colon \Psi_i(U_i \cap U_j) \to \Psi_j(U_i \cap U_j)$$

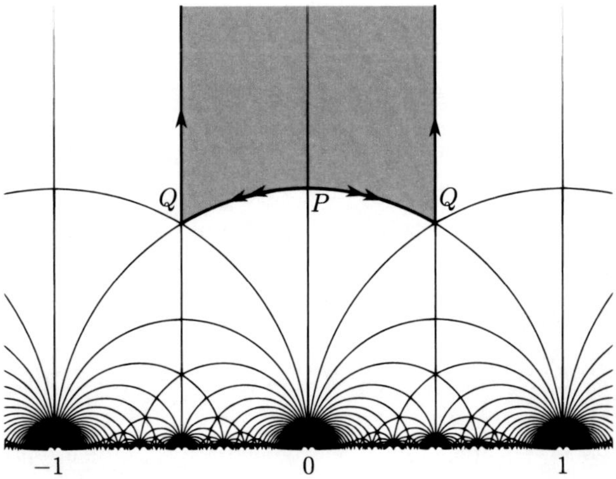

Figure 2.3: Moduli space of marked Euclidean structures up to homothety on the torus.

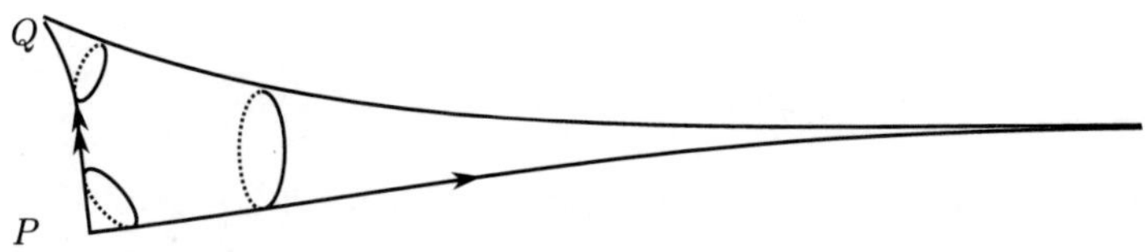

Figure 2.4: Moduli spaces of unmarked Euclidean structures up to homothety on the torus.

are restrictions of projective maps.

A given surface may have more than one projective structure. The data of a topological surface with a fixed real projective structure is called a *real projective surface.*

The following definitions are standard, for explicit details, see [48]. Let $(\widetilde{S}, \pi)$ be a universal cover of a real projective surface S. By choosing a base point $\widetilde{p} \in \widetilde{S}$ and a (subset of a) coordinate chart $\Psi : U \to V \subset \mathbb{RP}^2$ with $\pi(\widetilde{p}) \in U$, there is a real analytic diffeomorphism $\Psi \circ \pi : \pi^{-1}(U) \to V$. Analytically extending the domain of this map to all of $\widetilde{S}$, one obtains a *developing map*

$$\text{dev} : \widetilde{S} \to \mathbb{RP}^2.$$

As the fundamental group $\pi_1(S) = \pi_1(S, \pi(\widetilde{p}))$ acts on $\widetilde{S}$ by deck transformations, there is a unique *holonomy representation*

$$\text{hol}\colon\ \pi_1(S) \to \text{SL}(3, \mathbb{R}),$$

such that

$$\mathrm{hol}(\gamma) \cdot (\mathrm{dev}(\widetilde{x})) = \mathrm{dev}(\gamma \cdot \widetilde{x}), \qquad \forall\, \widetilde{x} \in \widetilde{S} \text{ and } \forall\, \gamma \in \pi_1(S).$$

The group $\Gamma = \mathrm{hol}(\pi_1(S))$ leaves $\Omega = \mathrm{dev}(\widetilde{S})$ invariant. Hence $\Gamma \leqslant \mathrm{SL}(\Omega)$. In particular, the geometric structure on S determined by the choice of developing map is modelled on the subgeometry $(\mathrm{SL}(\Omega), \Omega)$ of $(\mathrm{SL}(3,\mathbb{R}), \mathbb{RP}^2)$.

Assuming the universal cover $(\widetilde{S}, \pi)$ is fixed, a real projective structure determines a developing map and a holonomy representation up to the action of $\mathrm{SL}(3,\mathbb{R})$:

$$T \cdot (\mathrm{dev}, \mathrm{hol}) = (T \cdot \mathrm{dev}, T \cdot \mathrm{hol} \cdot T^{-1}), \qquad A \in \mathrm{SL}(3,\mathbb{R}).$$

Vice-versa, an $\mathrm{SL}(3,\mathbb{R})$–equivalence class of pairs $[\mathrm{dev}, \mathrm{hol}]$ uniquely determines a real projective structure. For that reason, henceforth we will always assume that a preferred universal cover $(\widetilde{S}, \pi)$ is fixed.

The choice of a representative in $[\mathrm{dev}, \mathrm{hol}]$ that leads to a marking will be discussed in detail in §2.2.6.

2.2.3 Representation spaces

Our holonomy representations are homomorphisms of $\pi_1(S)$ into $\mathrm{SL}(3,\mathbb{R})$, and a natural question that arises is which homomorphisms appear in this way, and how local properties of spaces of representations translate to local properties of spaces of geometric structures. The reader is referred to [19, §3] for a complete treatment of the case of a compact manifold, and to [11, §1] for the general case.

The set of homomorphisms $\mathrm{Hom}(\pi_1(S), \mathrm{SL}(3,\mathbb{R}))$ can be viewed as an algebraic set, which is well defined up to isomorphism. The construction is as follows. Given a finite presentation (with a symmetric generating set) for $\pi_1(S)$, a representation is uniquely determined by the images of the generators in $\mathrm{SL}(3,\mathbb{R})$. If there are n generators, this is a point in $\mathrm{SL}(3,\mathbb{R})^n \subset \mathbb{R}^{9n}$. This point satisfies algebraic equations arising from the relators in the presentation. Conversely, any point in $\mathbb{R}^{9n}$ that satisfies these equations, and the additional equations setting all determinants equal to one, gives rise to a representation of $\pi_1(S)$ into $\mathrm{SL}(3,\mathbb{R})$. Choosing a different generating set leads to an algebraic isomorphism between varieties, since the new generators are words in the old generators and we have chosen a symmetric generating set. An algebraic isomorphism preserves regularity of local rings, so the notion of a representation in $\mathrm{Hom}(\pi_1(S), \mathrm{SL}(3,\mathbb{R}))$ being a singular point or a smooth point of the variety is independent of the realisation of $\mathrm{Hom}(\pi_1(S), \mathrm{SL}(3,\mathbb{R}))$ as an affine algebraic set (cf. [26, Ch. 1 §5]).

The action of $\mathrm{SL}(3,\mathbb{R})$ on $\mathrm{Hom}(\pi_1(S), \mathrm{SL}(3,\mathbb{R}))$ by conjugation is algebraic. There is an open subset $\mathrm{Hom}(\pi_1(S), \mathrm{SL}(3,\mathbb{R}))^{st} \subset \mathrm{Hom}(\pi_1(S), \mathrm{SL}(3,\mathbb{R}))$ which consists of those orbits of the action on which the action of $\mathrm{SL}(3,\mathbb{R})$ is proper. These orbits are called *stable.* The $\mathrm{SL}(3,\mathbb{R})$–*character variety* of S is the closure of the space of stable orbits and it is denoted

$$\mathfrak{X}_3(S) := \mathrm{Hom}(\pi_1(S), \mathrm{SL}(3,\mathbb{R})) /\!\!/ \mathrm{SL}(3,\mathbb{R}).$$

This is also called the algebro-geometric quotient of $\mathrm{Hom}(\pi_1(S), \mathrm{SL}(3,\mathbb{R}))$ under the conjugation action of $\mathrm{SL}(3,\mathbb{R})$. The space $\mathrm{Hom}(\pi_1(S), \mathrm{SL}(3,\mathbb{R}))^{st}/\,\mathrm{SL}(3,\mathbb{R})$ has a natural real analytic structure from this set-up, and the (conjugacy classes of) representations that are of interest in these notes are contained in it.

2.2.4 Some projective structures on the torus

It is tempting to think of the developing map as determined by the holonomy representation (at least up to projective equivalence). Sullivan and Thurston [46] gave the following examples of inequivalent projective structures on the torus.

Consider the following 1–parameter family in $\mathrm{SL}(3,\mathbb{R})$.

$$A_t = \begin{pmatrix} e^{-3t} & 0 & 0 \\ 0 & e^t & 0 \\ 0 & 0 & e^{2t} \end{pmatrix}.$$

This defines a flow on $\mathbb{RP}^2$ with one source, one sink and one saddle, corresponding to the three standard basis vectors. Choose any closed curve γ transverse to the flow lines. Flowing this curve from time $t = 0$ to $t = 1$ sweeps out an immersed annulus in $\mathbb{RP}^2$. The two boundary components can be identified by the map A_1. The identification space is a projective structure on the torus $S^1 \times S^1$, whose developing map factors through the infinite annulus $\mathbb{R}^2 \to S^1 \times \mathbb{R} \to \mathbb{RP}^2$. The associated holonomy representation $\pi_1(S^1 \times S^1)$ has the infinite cyclic image $\langle A_1 \rangle$.

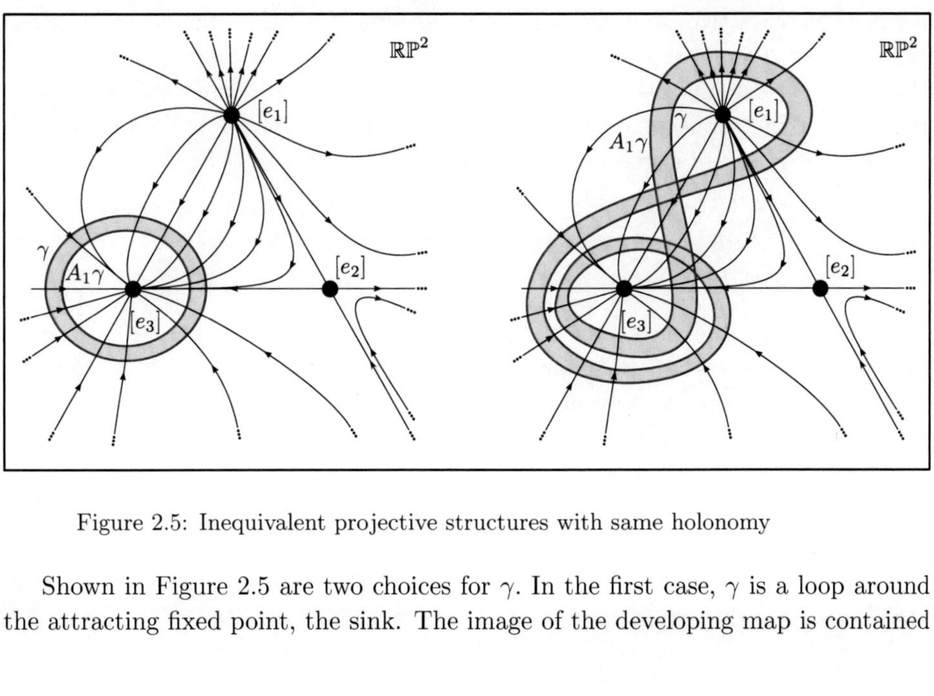

Figure 2.5: Inequivalent projective structures with same holonomy

Shown in Figure 2.5 are two choices for γ. In the first case, γ is a loop around the attracting fixed point, the sink. The image of the developing map is contained in the affine patch disjoint from the line passing through the source and the saddle. Moreover, this is an affine structure on the torus since A_1 is conjugate into the affine group Aff(2). It is not Euclidean, since A_1 is not conjugate into Euc(2).

In the second case, the image of the developing map is the whole complement of the source, the sink and the saddle. In particular, it is not contained in an affine patch and therefore not an affine structure on the torus.

2.2.5 Completeness and convexity

The previous examples are meant as a cautionary tale. The situation that we are mainly interested in is where the developing map is a homeomorphism onto its image and the holonomy representation is injective.

We say that a (G, X)–structure is *complete* if some developing map is a covering map. In particular, if X is simply connected, then the developing map is a homeomorphism.

Exercise 14. *Show that if some developing map is a covering map, then any developing map is a covering map.*

Let $\Omega \subset \mathbb{RP}^2$ be an open domain. We say that Ω is

- *convex* if $\Omega \cap \ell$ is connected, for all lines ℓ in $\mathbb{RP}^2$;
- *properly convex* if Ω is convex and $\overline{\Omega}$ is contained in an affine patch;
- *strictly convex* if Ω is properly convex and $\partial\overline{\Omega}$ intersects each line in at most two points;
- *round* if Ω is strictly convex and $\partial\overline{\Omega}$ is C^1.

We say that a real projective structure on S is *properly convex* if its developing map dev is an embedding onto a properly convex domain. It is common usage to omit the adjective *properly* and simply say *convex*. If, in addition, the image of dev is strictly convex, we say that the structure on S is *strictly convex*. Note that $\mathrm{SL}(3, \mathbb{R})$ preserves the above notions of convexity so the notion of a convex projective structure is well-defined. Moreover, the fact that dev is an embedding implies that hol is an isomorphism onto a discrete subgroup of $\mathrm{SL}(3, \mathbb{R})$.

Proposition 2.2.1. *If $\Gamma \leqslant \mathrm{SL}(3, \mathbb{R})$ is a discrete subgroup whose action on $\mathbb{RP}^2$ restricts to a free action on a properly convex domain Ω, then Ω/Γ is a surface with convex projective structure.*

Proof. It suffices to show that the action is properly discontinuous, i.e. that for every compact set $K \subset \Omega$ the intersection $K \cap gK$ is non-empty for at most finitely many $g \in \Gamma$. The reader will find pleasure in expanding the following paragraph using the material on isometries and the Hilbert metric in §2.3.

Suppose $K \subset \Omega$ is compact and there is a sequence of distinct elements $\gamma_i \in \Gamma$ with $K \cap \gamma_i K \neq \phi$. Choose a projective basis $\mathcal{B} = \{x_0, \cdots, x_n\} \subset \Omega$ with $x_0 \in K$. After taking a subsequence we may assume $\gamma_i \mathcal{B}$ converges to a subset of Ω. The sequence $\delta_i = \gamma_{i+1}^{-1}\gamma_i \in \Gamma$ has the property that $\delta_i \mathcal{B} \to \mathcal{B}$ because δ_i is an *isometry*. By Exercise 13, this implies that δ_i converges to the identity. ■

2.2.6 Marked structures

The previous result shows that surfaces with properly convex projective structures arise as quotient spaces of discrete group actions on properly convex open domains. Conversely, a properly convex projective structure on a surface gives rise to such a group action. This allows us to pass (up to projective equivalence) from a pair consisting of a developing map and a holonomy representation to a pair consisting of an open domain and a discrete group, and back again.

To make this precise, given a pair (dev, hol) for a properly convex projective structure on the surface S, we let $\Omega = \text{dev}(\widetilde{S})$ and $\Gamma = \text{hol}(\pi_1(S))$. Conversely, given a pair (Ω, Γ), we have a natural identification $S = \Omega/\Gamma$, $\widetilde{S} = \Omega$ and $\pi_1(S) = \Gamma$, from which we obtain the developing map and the holonomy by forgetting the extra structure on the left-hand side of the equalities.

We now wish to introduce the notion of a marking. Instead of doing it via marked discrete groups, as done in the example of the Euclidean torus above, we follow the classical approach of using maps from a reference surface. Hence suppose S is a topologically finite surface, i.e. the complement in a closed surface of a finite (possibly empty) union of points.

A *marked properly convex projective structure* on S is a triple (Ω, Γ, f), where

- $\Omega \subset \mathbb{RP}^2$ is a properly convex domain,
- Γ is a discrete (torsion-free) subgroup of $\text{PGL}(\Omega)$,
- $f\colon S \to \Omega/\Gamma$ is a real analytic diffeomorphism.

There is a natural notion of *equivalence* between marked properly convex projective structures. Two marked properly convex projective structures $(\Omega_0, \Gamma_0, f_0)$ and $(\Omega_1, \Gamma_1, f_1)$ are *equivalent* if there is $T \in \text{SL}(3, \mathbb{R})$ such that

$$T \cdot \Omega_0 = \Omega_1, \qquad T \cdot \Gamma_0 \cdot T^{-1} = \Gamma_1, \qquad \text{and} \qquad f_1 \text{ is isotopic to } T^* \cdot f_0,$$

where T^* is the map $\Omega_0/\Gamma_0 \to \Omega_1/\Gamma_1$ induced by T. On the level of developing maps and holonomies, this translates into the conjugacy $T \cdot \text{hol}_0 \cdot T^{-1} = \text{hol}_1$ and an isotopy between $T \cdot \text{dev}_0$ and dev_1.

This notion of equivalence is also the reason why we do not need to specify base points for fundamental groups — a different choice of base points results in conjugation within the groups $\pi_1(S)$ and Γ respectively, and an appropriate isomorphism $\text{hol} = f_*\colon \pi_1(S) \to \Gamma$.

The *moduli space of all marked properly convex projective structures* on the surface S is denoted $\mathcal{T}_3(S)$. As stated in the introduction, we view a moduli space as a space of objects, and seek suitable parametrisations of this moduli space. Having fixed a universal cover of S to take care of the marking, we can view elements of $\mathcal{T}_3(S)$ as pairs consisting of a developing map and a holonomy representation. The space $\mathcal{T}_3(S)$ has the following natural topology. The set of developing maps can be viewed as a subspace of the set $C^\infty_w(\widetilde{S}, \mathbb{RP}^2)$ of all smooth maps with the weak topology (see Hirsch [28, page 35]). We imbue $\mathcal{T}_3(S)$ with this topology (thus ignoring the holonomy representation). With respect to this, two developing maps for S are close if they are close on a large compact set in $\widetilde{S}$.

Recall that $\mathfrak{X}_3(S) := \mathrm{Hom}(\pi_1(S), \mathrm{SL}(3,\mathbb{R})) /\!\!/ \mathrm{SL}(3,\mathbb{R})$ is the $\mathrm{SL}(3,\mathbb{R})$–character variety for S. The *monodromy operator* is the natural map

$$\begin{aligned} \mu : \mathcal{T}_3(S) &\to \mathfrak{X}_3(S), \\ (\mathrm{dev}, \mathrm{hol}) &\mapsto [\mathrm{hol}]. \end{aligned}$$

A fact, often referred to as the Thurston-Lok principle [36], is the *deformation theorem*, which states that this map is continuous and open for very general reasons (see [19, §3]) if S is compact. A relative version for the non-compact case is given in [11, §1]. We will not make use of the deformation theorem, but rather recover a shadow of it in our setting.

2.2.7 Classification of properly convex automorphisms

Let $T \in \mathrm{SL}(3,\mathbb{R})$. Then T has at least one real eigenvalue. There are four possible real normal forms for T:

$$J_1 = \begin{pmatrix} \lambda_1 & 0 & 0 \\ 0 & \lambda_2 & 0 \\ 0 & 0 & \frac{1}{\lambda_1\lambda_2} \end{pmatrix}, \quad J_2 = \begin{pmatrix} \lambda_1^2 & 0 & 0 \\ 0 & \frac{1}{\lambda_1} & 1 \\ 0 & 0 & \frac{1}{\lambda_1} \end{pmatrix}, \quad J_3 = \begin{pmatrix} 1 & 1 & 0 \\ 0 & 1 & 1 \\ 0 & 0 & 1 \end{pmatrix},$$

and

$$J_4 = \begin{pmatrix} \lambda_1^2 & 0 & 0 \\ 0 & \frac{1}{\lambda_1}\cos\theta & -\frac{1}{\lambda_1}\sin\theta \\ 0 & \frac{1}{\lambda_1}\sin\theta & \frac{1}{\lambda_1}\cos\theta \end{pmatrix}.$$

To analyse some features of the action of J_k, it is useful to consider the action planes of the form $P_{ij} = \langle e_i, e_j \rangle$ spanned by two standard basis vectors, and to study forward orbits of the action of J_k. The following lemma follows the exposition of Marquis [38].

Lemma 2.2.2. *Suppose Ω is a properly convex domain and $T \in \mathrm{SL}(\Omega)$. Then T belongs one of the following mutually exclusive families:*

1. *The family of* totally hyperbolic transformations, *which are conjugate to matrices of the form:*

$$\begin{pmatrix} \lambda_1 & 0 & 0 \\ 0 & \lambda_2 & 0 \\ 0 & 0 & \lambda_3 \end{pmatrix}, \quad \textit{where } \lambda_1 > \lambda_2 > \lambda_3 > 0 \textit{ and } \lambda_1\lambda_2\lambda_3 = 1.$$

2. *The family of* planar transformations, *which are conjugate to matrices of the*

form:

$$\begin{pmatrix} \lambda_1 & 0 & 0 \\ 0 & \lambda_2 & 0 \\ 0 & 0 & \lambda_2 \end{pmatrix}, \text{ where } \lambda_1, \lambda_2 > 0, \lambda_1\lambda_2^2 = 1 \text{ and } \lambda_1, \lambda_2 \neq 1.$$

3. *The family of* quasi-hyperbolic transformations, *which are conjugate to matrices of the form:*

$$\begin{pmatrix} \lambda_1 & 0 & 0 \\ 0 & \lambda_2 & 1 \\ 0 & 0 & \lambda_2 \end{pmatrix}, \text{ where } \lambda_1, \lambda_2 > 0, \lambda_1\lambda_2^2 = 1 \text{ and } \lambda_1, \lambda_2 \neq 1.$$

4. *The family of* parabolic transformations, *which are conjugate to matrices of the form:*

$$\begin{pmatrix} 1 & 1 & 0 \\ 0 & 1 & 1 \\ 0 & 0 & 1 \end{pmatrix}.$$

5. *The family of* elliptic transformations, *which are conjugate to matrices of the form:*

$$\begin{pmatrix} 1 & 0 & 0 \\ 0 & \cos\theta & -\sin\theta \\ 0 & \sin\theta & \cos\theta \end{pmatrix}, \text{ where } 0 < \theta < 2\pi.$$

6. *The family consisting only of the* identity transformation.

Proof. First suppose that not all eigenvalues of T are real. Then T has the normal form J_4. Now, the matrix J_4 fixes $[e_1] \in \mathbb{RP}^2$. If $\lambda_1 \neq 1$, then J_4 acts as a spiral expansion or as a spiral contraction on the plane $P_{23} = \langle e_2, e_3 \rangle$. It is now not difficult to see that J_4 only leaves a properly convex domain invariant if $\lambda_1 = 1$.

Hence suppose that all eigenvalues of T are real.

Suppose that all eigenvalues of T are contained in $\{\pm 1\}$. In particular, all eigenvalues of T^2 are equal to 1. We show that T^2 cannot have real normal form J_2. This follows since otherwise the projectivisation of $P_{12} = \langle e_1, e_2 \rangle$ is a line fixed pointwise by J_2. Given any other projective line ℓ through $[e_2]$, the action of J_2 on $\ell \setminus [e_2]$ is by translation. Hence J_2 does not preserve a properly convex domain. In particular, this is also true for T^2. This shows that, if all eigenvalues of T are contained in $\{\pm 1\}$, then T^2 is conjugate to either a parabolic or the identity. It now follows that T is either a parabolic, or a reflection (conjugate to the elliptic $\text{diago}(1, -1, -1)$), or the identity.

The last case is that all eigenvalues are real, but at least one (and hence at least two) of the eigenvalues is not contained in $\{\pm 1\}$. We may assume that $|\lambda_1| > 1$, $1 > |\lambda_3|$ and $|\lambda_1| \geqslant |\lambda_2| \geqslant |\lambda_3|$. Since there are at least two distinct eigenvalues, there is a point $p \in \Omega$ that is not contained in the projectivisation of an eigenspace of T. The projective line through p and Tp is divided into two segments with these endpoints, and by convexity of Ω, one of these segments is contained in Ω. Denote this segment by s. If $\lambda_1 < 0$, then the sequence of segments $T^{2n}s$ converges to a line in $\mathbb{RP}^2$. This contradicts the proper convexity of Ω. Hence $\lambda_1 > 0$. Applying the same argument to T^{-1} gives $\lambda_3 > 0$. This also forces $\lambda_2 > 0$. This completes the proof of the lemma. ■

Exercise 15. *Suppose Ω is a properly convex domain and $T \in \mathrm{SL}(\Omega)$ is elliptic. Then T has a fixed point in Ω.*

Exercise 16. *Suppose Ω is a properly convex domain and $T \in \mathrm{SL}(\Omega)$ is planar. Then Ω is a triangle with one vertex corresponding to the isolated fixed point of T and the side opposite to this vertex is contained in the line fixed pointwise by T.*

2.2.8 Classification of ends

We recall the notation from the introduction: S_g denotes a closed, orientable surface of finite genus g; $D_1, \ldots, D_n$ are open discs on S_g with pairwise disjoint closures; and $p_k \in D_k$. Then $S_{g,n} = S_g \backslash \{p_1, \ldots, p_n\}$ is a punctured surface and $\mathcal{E}_k = \overline{D_k} \backslash \{p_k\}$ is an *end* of $S_{g,n}$. A *compact core* of $S_{g,n}$ is $S^c_{g,n} = S_g \backslash \cup D_k$. We also write $S_g = S_{g,0}$. We assume throughout that $S_{g,n}$ has negative Euler characteristic.

Furthermore, we fix a marked convex projective structure (Ω, Γ, f) on $S = S_g$, with $\Omega = \mathrm{dev}(\widetilde{S})$ and $\Gamma = \mathrm{hol}(\pi_1(S))$. Recall that dev is a homeomorphism onto its image and hol is an isomorphism. For each end $\mathcal{E}_k$, we identify $\pi_1(\mathcal{E}_k)$ with its image in $\pi_1(S)$ and call it a *peripheral subgroup* of $\pi_1(S)$. Similarly, $\Gamma_k := \mathrm{hol}(\pi_1(\mathcal{E}_k))$ is called a *peripheral subgroup* of Γ.

We first note that no element of Γ is elliptic or planar. Indeed, if there is an elliptic element, then since the action is properly discontinuous, this would have to be of finite order. But $\pi_1(S)$ (and hence Γ) is torsion free. If Γ contains a planar element, then Ω is a triangle. It is not difficult to see that in this case $\mathrm{SL}(\Omega)$ is the semi-direct product an abelian group (namely the group of all elements with the vertices of the triangle corresponding to eigenvectors) with the symmetric group on the three vertices (see §2.3.2). Thus, $\Gamma \leqslant \mathrm{SL}(\Omega)$ contains a finite index subgroup that is abelian. This contradicts the fact that Γ is isomorphic to the fundamental group of an orientable surface with negative Euler characteristic. It follows that a non-trivial element of Γ is either totally hyperbolic, quasi-hyperbolic or parabolic.

Let γ be a generator of the peripheral subgroup $\pi_1(\mathcal{E})$, of an end $\mathcal{E}$. Associated with this is an embedding of the universal cover of the end into the universal cover of S that is left invariant under γ. We identify this preferred embedding as a subset, $\widetilde{\mathcal{E}} \subset \widetilde{S}$. Now $\mathrm{hol}(\gamma)$ preserves the convex domain $\Omega = \mathrm{dev}(\widetilde{S})$, and furthermore, $\mathrm{hol}(\gamma)$ preserves $\mathrm{dev}(\widetilde{\mathcal{E}}) \subset \Omega$ by equivariance. According to the above discussion,

$\mathrm{hol}(\gamma) \in \mathrm{SL}(3,\mathbb{R})$ is conjugate to one of the following three Jordan forms:

$$J_1 = \begin{pmatrix} \lambda_1 & 0 & 0 \\ 0 & \lambda_2 & 0 \\ 0 & 0 & \frac{1}{\lambda_1\lambda_2} \end{pmatrix}, \qquad J_2 = \begin{pmatrix} \lambda_1^2 & 0 & 0 \\ 0 & \frac{1}{\lambda_1} & 1 \\ 0 & 0 & \frac{1}{\lambda_1} \end{pmatrix}, \qquad J_3 = \begin{pmatrix} 1 & 1 & 0 \\ 0 & 1 & 1 \\ 0 & 0 & 1 \end{pmatrix}.$$

$$\lambda_1 > \lambda_2 > \frac{1}{\lambda_1\lambda_2}, \qquad \lambda_1 \neq 1.$$

We now discuss the dynamics of these transformations in a little more detail. This will be further developed in §4.2.

In the first case, the transformation $\mathrm{hol}(\gamma) \sim J_1$ is totally hyperbolic. It fixes three pairwise distinct points of $\mathbb{RP}^2$ and preserves the lines through them (see Figure 2.6). The *attractive* (resp. *repelling*) point is the one corresponding to the largest (resp. smallest) eigenvalue. The remaining point is the *saddle* point. In this case, we refer to $\mathcal{E}$ as a *hyperbolic end.*

In the second case, $\mathrm{hol}(\gamma) \sim J_2$ is quasi-hyperbolic and fixes only two distinct points of $\mathbb{RP}^2$, preserves the line through them and acts as a unipotent transformation on a second line. In this case we say that $\mathcal{E}$ is a *special end.*

In the last case, $\mathrm{hol}(\gamma) \sim J_3$ is parabolic, it fixes a unique point and preserves a unique line through it. Hence $\mathcal{E}$ is a *cusp.*

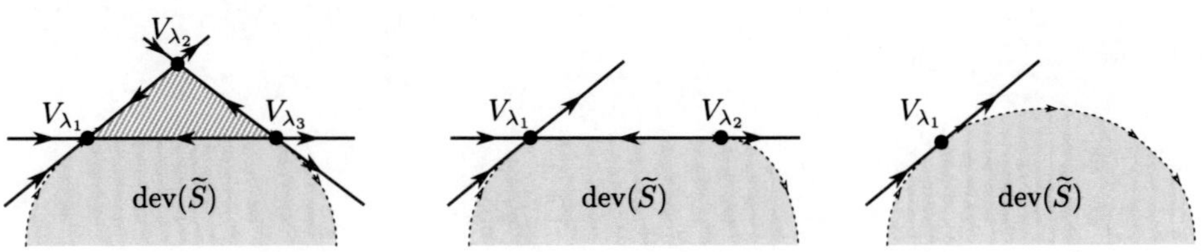

Figure 2.6: Actions of a totally hyperbolic, quasi-hyperbolic and parabolic peripheral element, respectively.

The action of $\mathrm{hol}(\pi_1(\mathcal{E}))$ preserves Ω, so $\partial\overline{\Omega}$ always contains the attracting and repelling fixed points of the peripheral element $\mathrm{hol}(\gamma)$ (whether or not they are distinct). In particular, when $\mathcal{E}$ is a hyperbolic or special end, the interior of a segment between the the attracting and repelling fixed points is either contained in Ω or $\partial\overline{\Omega}$. Fix an affine patch $\mathcal{A}$ containing $\overline{\Omega}$. A hyperbolic end is *maximal* if Ω contains the interior of a triangle spanned in $\mathcal{A}$ by the three fixed points of $\mathrm{hol}(\gamma)$, and it is *minimal* if Ω is contained in the complement of this triangle. The triangle in question is shaded in the left image of Figure 2.6. Note that Ω cannot be equal to the interior of such a triangle since otherwise S would itself be an end, hence an annulus, contradicting our hypothesis on the topology of S. We define $\mathcal{T}_3^m(S) \subset \mathcal{T}_3(S)$ to be the subset of structures for which *every hyperbolic end is either maximal or minimal.*

2.3 The Hilbert metric

Let $\Omega \subset \mathbb{RP}^2$ be a properly convex open set contained in an affine patch endowed with some arbitrary Euclidean norm $\|\cdot\|$. The *Hilbert metric* d_Ω on Ω is

$$d_\Omega(a,b) := \log[x,a,b,y] = \log \frac{\|b-x\| \cdot \|a-y\|}{\|b-y\| \cdot \|a-x\|},$$

where $x, y \in \partial\Omega$ are the endpoints of a line segment in Ω containing a and b, such that a lies between x and b. The definition is worded to include the case $a = b$. See [5] and [12] for proofs that this is a metric using cross ratios, and [17] for a proof using the Kobayashi metric.

It is easy to verify that for any c lying on the segment between a and b, we have $d_\Omega(a,b) = d_\Omega(a,c) + d_\Omega(c,b)$. Whence every segment of a projective line in Ω is length minimising, and in the case that Ω is strictly convex, these are the only geodesics.

There is an interesting comparison principle. If $\Omega \subseteq \Omega'$ and $a, b \in \Omega$, then

$$d_\Omega(a,b) \geqslant d_{\Omega'}(a,b).$$

Projective transformations do not change the cross ratio, so the group $\mathrm{SL}(\Omega)$ of transformations preserving Ω is a subgroup of the group $\mathrm{Isom}(\Omega, d_\Omega)$ of isometries of the Hilbert metric. The inclusion $\mathrm{SL}(\Omega) \leqslant \mathrm{Isom}(\Omega, d_\Omega)$ is strict if and only if Ω is the interior of a triangle with respect to some affine patch (see [39] for a proof of this fact). The triple

$$(\Omega, d_\Omega, \mathrm{SL}(\Omega))$$

is called a *Hilbert geometry.*

Associated to $T \in \mathrm{SL}(\Omega)$ is its *translation length*

$$t(T) = \inf_{x\in\Omega} d_\Omega(x, Tx).$$

If T is elliptic, then $t(T) = 0$ is independent of Ω. It turns out that the translation length is indeed determined by the *algebra* of T and not the *geometry* of Ω. The following exercise is pleasant in dimension two (see Proposition 2.1 in [10] for the general case):

Exercise 17. *Suppose* $T \in \mathrm{SL}(\Omega)$, *and let* λ *and* μ *be the (possibly complex) eigenvalues of maximum and minimum modulus. Then* $t(T) = \log|\lambda/\mu|$.

The Hilbert metric descends to a metric on our quotient spaces via the following, more general, notions. Suppose that (X, d) is a metric space and G is a group of isometries of X. Then the quotient semi-metric $\overline{d}$ on X/G is given by:

$$\overline{d}(\overline{x}, \overline{y}) := \inf\{d(x, \gamma \cdot y) \mid \gamma \in G\},$$

where $x \mapsto \overline{x}$ is the quotient map $X \to X/G$. If the action of G on X is properly discontinuous, then $\overline{d}$ is a metric. If the action is properly discontinuous and free, then X/G is locally isometric to X. Details can be found in [48] and [4].

2.3.1 The Klein model of the hyperbolic plane

The *Klein model* or *projective model of the hyperbolic plane* $\mathbb{H}^2$ is the Hilbert geometry with domain the standard open unit ball $\mathbb{D}^2 \subset \mathbb{R}^2 \subset \mathbb{RP}^2$. We have:

$$\mathrm{SL}(\mathbb{D}^2) \cong \mathrm{PO}(1,2) = \{T \in \mathrm{GL}(3) \mid T^{\mathsf{T}} J T = J\}/\{\pm I\},$$

where $J = \operatorname{diago}(-1,1,1)$. The Hilbert metric gives *twice* the usual hyperbolic metric (information on adjusting the *curvature* is given in §2.3.3). More generally, when Ω is the interior of an ellipsoid, one obtains a Hilbert geometry isometric with this, since one can take any ellipsoid to a ball by an affine map.

More familiar models of the hyperbolic plane are the half-space model and the Poincaré disc model. In addition, there is a hyperboloid model and a hemisphere model. An excellent account of the relationships between these models is [6]. Of particular interest for these notes is that the transition maps between any two models are real analytic, so results obtained in one model can be transferred to another model.

2.3.2 The Hilbert metric on a simplex

The *Hex plane* is obtained by taking $\Omega = \Delta$, the interior of an open 2–simplex. Then $\mathrm{SL}(\Delta)$ is the semi-direct product of positive diagonal matrices of determinant one and permutations of the three vertices:

$$\mathrm{SL}(\Delta) \cong \{\operatorname{diago}(a,b,c) \mid a,b,c > 0, abc = 1\} \rtimes \mathrm{Sym}(3).$$

An excellent reference for this geometry is de la Harpe [12]. This Hilbert geometry is isometric to a normed vector space, where the unit ball is a regular hexagon. To make this explicit, let $(\mathbb{R}^2, \|\cdot\|_{hex})$ be the vector space with *Hex norm*:

$$\|(x,y)\|_{hex} := \frac{1}{2\sqrt{3}}(|2y| + |y + \sqrt{3}x| + |y - \sqrt{3}x|).$$

Choosing the vertices $a_1 = (1,0)$, $a_2 = (-1/2, \sqrt{3}/2)$ and $a_3 = (-1/2, -\sqrt{3}/2)$ of the hexagon, an isometry between (Δ, d_Δ) and $(\mathbb{R}^2, \|\cdot\|_{hex})$ is given by

$$[e^{t_1}e_1 + e^{t_2}e_2 + e^{t_3}e_3] = [e^{t_1} : e^{t_2} : e^{t_3}]^{\mathsf{T}} \mapsto t_1a_1 + t_2a_2 + t_3a_3.$$

Since the unit ball is *not strictly convex* geodesics are *not even locally unique.* The isometry $x \mapsto -x$ of $(\mathbb{R}^2, \|\cdot\|_{hex})$ does not correspond to a projective transformation of (Δ, d_Δ) since it is given by

$$[x : y : z]^{\mathsf{T}} \mapsto \left[\frac{1}{x} : \frac{1}{y} : \frac{1}{z}\right]^{\mathsf{T}}.$$

It turns out that

$$\mathrm{Isom}(\Delta, d_\Delta) \cong \{\mathrm{diago}(a,b,c) \mid a,b,c > 0, abc = 1\} \rtimes D_6,$$

where D_6 is the symmetry group of a regular hexagon (this dihedral group is also often denoted D_{12} to indicate its order rather than the object it acts on). Hence $\mathrm{SL}(\Delta)$ has index 2 in $\mathrm{Isom}(\Delta, d_\Delta)$.

Exercise 18. *If* $\square$ *is a quadrilateral in* $\mathbb{RP}^2$, *then* $\mathrm{SL}(\square) = \mathrm{Isom}(\square, d_\square) \cong D_4$, *where* D_4 *is the symmetry group of the square.*

More generally, de la Harpe [12] shows that for any convex n–gon P in $\mathbb{RP}^2$ with $n \geqslant 4$, both groups $\mathrm{SL}(P)$ and $\mathrm{Isom}(P, d_P)$ are finite, and, as mentioned above, it follows from [39] that $\mathrm{SL}(P) = \mathrm{Isom}(P, d_P)$.

The Hilbert metric on a simplex plays another special role:

Theorem 2.3.1 (Förtsch-Karlsson 2005, [16]). *The Hilbert geometry* (Ω, d_Ω) *is isometric to a normed vector space if and only if* Ω *is the interior of a simplex.*

2.3.3 The Hilbert metric as a Finsler or Riemannian metric

Busemann and Kelly [5] give a slightly more general definition of the Hilbert metric, which inserts a constant factor in the previous definition:

$$d_\Omega(a,b) := \frac{k}{2}\log[x,a,b,y] = \frac{k}{2}\log\frac{\|b-x\|\cdot\|a-y\|}{\|b-y\|\cdot\|a-x\|},$$

where $k > 0$. For any properly convex domain Ω, this gives a complete *Finsler metric* with the Finsler norm of $u \in \mathbb{R}^n$ at $a \in \Omega$ given by:

$$F_\Omega(a,u) = \frac{k\|u\|}{2}\left(\frac{1}{\|x-u^-\|} + \frac{1}{\|x-u^+\|}\right),$$

where u^- and u^+ are the intersection points of $\partial\Omega$ with the line based at a in direction u, such that u^+ is in the forwards direction. The length of a rectifiable path $\gamma\colon [0,1] \to \Omega$ is then given by:

$$L_\Omega(\gamma) = \int_{[0,1]} F_\Omega(\gamma(t), \gamma'(t))dt.$$

See Vernicos [49] for a proof of this fact. Some authors specialise to the case $k = 1$ (giving the hyperbolic metric in the Klein model) and others to the case $k = 2$ (eliminating constants in many expressions).

The differentiability class of the Finsler norm is precisely the differentiability class of $\partial\Omega$. In general, convexity implies that the boundary of Ω is C^1 almost everywhere with respect to Lebesgue measure on $\mathbb{RP}^n$. In order to apply techniques from Finsler geometry, one often needs to make additional hypotheses on the regularity of the boundary of Ω, generally making it at least of class C^2. See

Socié-Méthou [44] for a complete treatment of these and other aspects of the Finsler geometry of Hilbert geometries. In particular, the above metric (under the right regularity conditions) guarantees that the usual *Finsler curvature* is defined, making (Ω, d_Ω) into a Finsler manifold of constant negative curvature $-\frac{1}{k^2}$.

The relationship between Hilbert geometry and Riemannian geometry has been known for quite some time:

Theorem 2.3.2 (Kay 1967, [31]). *A Hilbert geometry (Ω, d_Ω) is Riemannian if and only if $\partial\overline{\Omega}$ is an ellipsoid. In this case, (Ω, d_Ω) is isometric with hyperbolic space.*

The up-shot of this discussion is that if one imposes extra regularity conditions on $\partial\overline{\Omega}$, then many technical obstacles will either not occur or may become easier to navigate as there are additional techniques from Finsler geometry at one's disposal.

2.3.4 Area

We define volume in a Hilbert geometry using the *Hausdorff measure*; an excellent reference for this is [45]. Here is a summary. For a set Y in a metric space (X, d), its *diameter* is

$$\operatorname{diam} Y = \sup\{d(x, y) \mid x, y \in Y\}.$$

Then the α–dimensional *Hausdorff measure* of a subset $Y \subset X$ is defined by first putting

$$\mathcal{H}^\delta_\alpha(Y) := \inf\Big\{\sum_k (\operatorname{diam} F_k)^\alpha \mid Y \subset \cup_{k=1}^\infty F_k, \operatorname{diam} F_k \leqslant \delta \text{ for all } k\Big\},$$

and then letting

$$\nu_\alpha(Y) := \lim_{\delta\to 0} \mathcal{H}^\delta_\alpha(Y).$$

This exists (allowing the value ∞) because the quantity $\mathcal{H}^\delta_\alpha(Y)$ increases as one decreases δ. For any set Y, there is a unique α such that

$$\nu_\beta(Y) = \begin{cases} 0 & \text{if } \beta > \alpha, \\ \infty & \text{if } \beta < \alpha. \end{cases}$$

This number α (which is not necessarily an integer) is called the *Hausdorff dimension* of Y.

Restricting our attention to a properly convex domain $\Omega \subset \mathbb{RP}^2$ with the Hilbert metric, the Hausdorff dimension of any open subset of Ω is 2. The 2–dimensional Hausdorff measure $\nu_\Omega = \nu_2$ on Ω depends only on d_Ω. It follows immediately from the definition of the Hilbert metric and the Hausdorff measure that $\nu_\Omega(Y) < \nu_{\Omega'}(Y)$, for all $Y \subset \Omega' \subset \Omega$ in $\mathbb{RP}^2$.

It is not difficult to see that $\nu_\Omega(\Omega) = \infty$. The measure ν_Ω is absolutely continuous with respect to the Lebesgue measure $\nu_\mathcal{L}$ (i.e. $\nu_\mathcal{L}(Y) = 0$ implies $\nu_\Omega(Y) = 0$ for any set Y that is measurable with respect to both measures) and vice versa.

Since the measure is defined in terms of the metric and $\mathrm{SL}(\Omega)$ acts by isometries, it preserves the measure, and one therefore obtains a well-defined measure on any projective manifold modelled on a Hilbert geometry. We can therefore distinguish between projective surfaces of finite and infinite area. The space of *properly convex projective structures on S having finite area* is denoted $\mathcal{T}_3^f(S)$. The relationship between the data of dev, hol and the question of whether or not a structure has finite area is elucidated in a result of Marquis [37]. We give a proof of this result, employing some results of Fock and Goncharov, in §6.1.

2.3.5 Benzécri's Theorem

The following classical result is due to Benzécri, and generalises to all dimensions. In this section, we follow the proof and discussion given in [11], as this gives an algorithm to transform a domain into *Benzécri position.* The main departure from [11] is the restriction to dimension two.

Theorem 2.3.3 (Benzécri 1960, [2]). *Suppose Ω is a properly convex open domain in $\mathbb{RP}^2$ and $p \in \Omega$. Then there are a projective transformation $T \in \mathrm{SL}(3,\mathbb{R})$ and an affine patch $\mathcal{A} = \mathbb{R}^2$ such that $T(\overline{\Omega}) \subset \mathcal{A}$, $T(p) = 0 \in \mathcal{A}$ and*

$$B(1) \subset T(\Omega) \subset B(14) \subset \mathcal{A},$$

where $B(r)$ is the closed ball of radius r in the affine patch $\mathcal{A}$ centered at 0.

The pair consisting of the projective transformation T and the affine patch $\mathcal{A}$ is called a *Benzécri chart for Ω centered at p* and the image $T(\Omega, p) \subset \mathcal{A}$ is called *Benzécri position.*

Proof. We are given Ω and $p \in \Omega$. Choose any projective line ℓ containing p. Then $s = \Omega \cap \ell$ is connected and we let $\partial\overline{\Omega} \cap \ell = \{p^+, p^-\}$.

We first show that we can choose coordinates on an affine patch $\mathcal{A}$ with $\overline{\Omega} \subset \mathcal{A}$ such that $s = (-1,1) \times \{0\}$ and $p = (0,0)$.

Choose an additional point $q \in \Omega \setminus s$. Since Ω is properly convex, there is a line ℓ' such that $\overline{\Omega} \subset \mathbb{RP}^2 \setminus \ell'$. The projective line ℓ' determines a plane $L' \subset \mathbb{R}^3$. Choose a normal vector n to this plane. We may now choose representatives $\tilde{p}, \tilde{p}^+, \tilde{p}^-, \tilde{q} \in \mathbb{R}^3$ of p, p^+, p^-, q such that $n \cdot \tilde{p} > 0, n \cdot \tilde{p}^+ > 0, n \cdot \tilde{p}^- > 0, n \cdot \tilde{q} > 0$. Moreover, since p lies between p^+ and p^- on the line ℓ, we may choose these representatives such that

$$\frac{1}{2}(\tilde{p}^+ + \tilde{p}^-) = \tilde{p}.$$

Let P be the affine plane in $\mathbb{R}^3$ containing $\tilde{p}^+$, $\tilde{p}^-$, and $\tilde{q}$, and P_0 be the plane parallel to P and containing the origin. Then the projectivisation $\mathbb{P}(P_0)$ of the plane P_0 is a projective line disjoint from $\overline{\Omega}$ in $\mathbb{RP}^2$, and the affine patch $\mathcal{A} = \mathbb{RP}^2 \setminus \mathbb{P}(P_0)$ has the desired affine coordinates by choosing $\tilde{p}$ as the origin and using the basis $(\tilde{p}^- - \tilde{p}, \tilde{q} - \tilde{p})$. We will refer to $\tilde{p}^- - \tilde{p}$ as the x_1– and to $\tilde{q} - \tilde{p}$ as the x_2–coordinate.

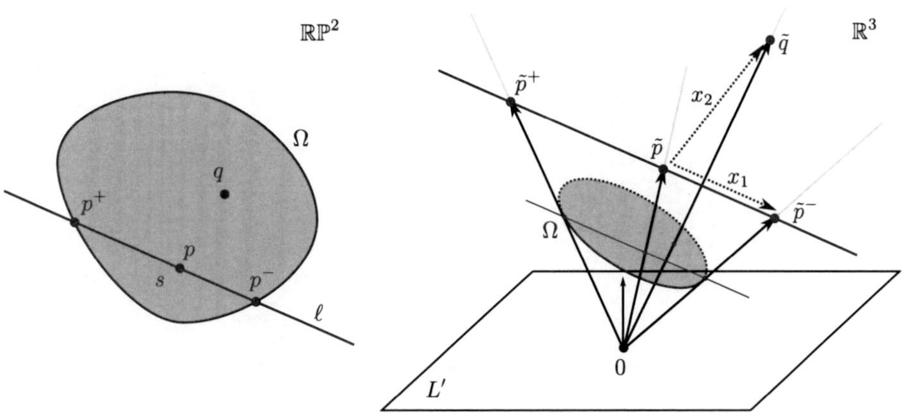

Figure 2.7: The affine coordinates for the Benzécri chart.

This completes the construction of the affine patch, hence of a Benzécri chart. To obtain a Benzécri position in this chart, we now apply projective maps that leave this patch invariant and fix $s = (-1, 1) \times \{0\}$ pointwise. Even though Ω will be moved by projective transformations, we continue to use the same symbol for the moved domain.

First, by applying a dilation in the x_2–coordinate we may arrange that $\Omega \subset \mathbb{R} \times [-1, 1]$ and, by possibly changing the positive direction along the x_2–axis, we may assume that $\overline{\Omega} \cap (\mathbb{R} \times \{1\}) \neq \varnothing$. We may then apply a shear so that $z = (0, 1) \in \overline{\Omega}$.

There is a 1–parameter family of projective maps that fixes s and z. With respect to the ordered basis $(\tilde{p}, \tilde{p}^+, (0, 1, 1)^\intercal)$ this family consists of the diagonal matrices $\operatorname{diago}(1, 1, \lambda)$, where $\lambda \in \mathbb{R}_{>0}$. This 1–parameter family fixes the line $\{x_2 = 0\}$, preserves the line $\{x_2 = 1\}$, and permutes the remaining lines $\{x_2 = \text{constant}\}$. Hence there is an element T_0 in this family with the property that

$$T_0(\Omega) \subset \mathbb{R} \times [-1, 1] \quad \text{and} \quad \overline{T_0(\Omega)} \cap (\mathbb{R} \times \{-1\}) \neq \varnothing.$$

By construction, $z \in \overline{T_0(\Omega)} \cap (\mathbb{R} \times \{1\})$. According to our conventions, we will omit T_0 from the notation and denote the resulting domain by Ω. We now show that this domain contains the interior of the ball $B(1/\sqrt{10})$ and that it is contained in the interior of the ball $B(\sqrt{17})$.

By convexity, the set $\overline{\Omega}$ contains the triangle with vertices $(-1, 0)$, $(0, 1)$ and $(1, 0)$. It also contains a point w on $\mathbb{R} \times \{-1\}$. Since $(-1, 0)$, $(0, 1)$ and $(1, 0)$ all lie on $\partial\overline{\Omega}$, it follows that $w \in [-2, 2] \times \{-1\}$. In particular, the ball $B(1/\sqrt{10})$ is contained in $\overline{\Omega}$ for all possible positions of w.

The line segment s' from w to any point in $\overline{\Omega}$ is contained in $\overline{\Omega}$. Using the fact that $\overline{\Omega} \cap (\mathbb{R} \times \{0\}) = [-1, 1] \times \{0\}$, and $\overline{\Omega} \subset \mathbb{R} \times [-1, 1]$, it follows that s' is

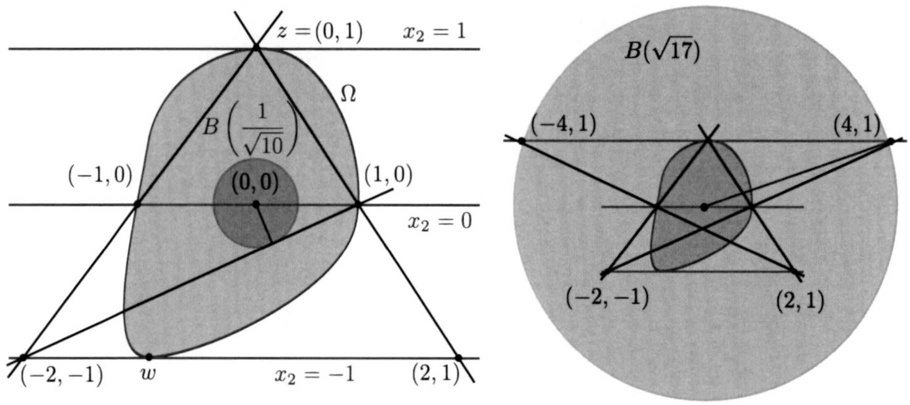

Figure 2.8: Balls in the Benzécri chart.

contained in a line segment between w and a point on $[-4,4]\times\{1\}$. In particular, $\overline{\Omega}$ is contained in the union of two convex quadrilaterals; one having vertices $(-4,1)$, $(4,1)$, $(1,0)$, $(-1,0)$ and the other having vertices $(-2,-1)$, $(2,-1)$, $(1,0)$, $(-1,0)$. This is contained in the closed ball $B(\sqrt{17})$.

The statement of the theorem now follows by dilating in all directions by $\sqrt{11}$:

$$B(1)\subset B\left(\frac{\sqrt{11}}{\sqrt{10}}\right)\subseteq\overline{\Omega}\subseteq B(\sqrt{187})\subset B(14).$$

It thus follows that $B(1)\subset\Omega\subset B(14)$. ■

Let $\mathcal{S}$ be the set of all $\Omega\subset\mathbb{RP}^2$, where Ω is compact, properly convex and with non-empty interior. Equip this set with the Hausdorff topology. Let $\mathcal{S}_*\subset\mathcal{S}\times\mathbb{RP}^2$ be the set of all pairs (Ω,p) with p in the interior of Ω with the subspace topology of the product topology. There is an action of $T\in\mathrm{SL}(3,\mathbb{R})$ on $\mathcal{S}_*$ given by $T(\Omega,p)=(T\Omega,Tp)$. The quotient of $\mathcal{S}_*$ by this action is given the quotient topology and denoted $\mathcal{Q}$.

Corollary 2.3.4 (Benzécri's compactness theorem [2]). *$\mathcal{Q}$ is compact.*

It follows that there is a compact set of preferred charts centered on a point in a properly convex surface. Different preferred charts give Euclidean coordinates around p which vary in a compact family independent of the surface. A Benzécri chart for (Ω,p) determines a Euclidean metric d_{euc} on Ω induced from the affine patch, a hyperbolic metric on Ω arising from the Hilbert metric from $B(14)$, and there is also the Hilbert metric d_Ω. In particular, distances and areas in a ball of d_Ω–radius r around p can be estimated in terms of the Euclidean or hyperbolic distances or areas *independent* of Ω. An application of this is given in the proof of Lemma 6.1.1.

Chapter 3

Ratios and Configurations of Flags

This section is devoted to the development of some foundational tools: triple ratios, cross ratios, quadruple ratios and configurations of flags. Most of the definitions and conventions follow Fock and Goncharov [15].

3.1 Flags

Recall from §2.1.7 that the analytic homomorphism $\mathrm{GL}(3,\mathbb{R}) \to \mathrm{SL}(3,\mathbb{R})$ defined by

$$A \mapsto (\det(A))^{-1/3}A$$

descends to an isomorphism $\mathrm{PGL}(3,\mathbb{R}) \to \mathrm{SL}(3,\mathbb{R})$. We work with $\mathrm{SL}(3,\mathbb{R})$, which allows us to talk about eigenvalues of maps, and it will be clear from context whether an element of $\mathrm{SL}(3,\mathbb{R})$ acts on $\mathbb{RP}^2$ or on $\mathbb{R}^3$. We also remark that the action on $\mathbb{R}^3$ is by orientation preserving maps.

A *flag* $\mathcal{F}_i := (V_i, \eta_i)$ of $\mathbb{RP}^2$ is a pair consisting of a point $V \in \mathbb{RP}^2$ and a line $\eta \subset \mathbb{RP}^2$ passing through V. An m–tuple of flags $\{\mathcal{F}_1, \dots, \mathcal{F}_m\}$ is in *general position* if all of the following conditions are satisfied:

- no three points are collinear;
- no three lines are concurrent;
- $\eta_i(V_j) = 0 \iff i = j$.

Henceforth, we will denote by $(\!(\mathcal{F}_1, \dots, \mathcal{F}_m)\!)$ a cyclically ordered m–tuple of flags, as opposed to an ordered m–tuple $(\mathcal{F}_1, \dots, \mathcal{F}_m)$. The group of projective transformations acts on the space of flags via:

$$T \cdot \mathcal{F}_i := (T \cdot V_i, T \cdot \eta_i), \qquad \text{for all } T \in \mathrm{SL}(3,\mathbb{R}).$$

The action naturally extends to m–tuples, ordered m–tuples and cyclically ordered m–tuples of flags. Furthermore, the action preserves the respective subspaces of flags in general position.

Lemma 3.1.1. *The action of* $\mathrm{SL}(3,\mathbb{R})$ *on the space of ordered* m*–tuples of flags in general position is free if and only if* $m \geqslant 3$.

Proof. The cases $m = 1, 2$ are straight forward.

Suppose $m \geqslant 3$ and let $\mathfrak{F} = (\mathcal{F}_1, \dots, \mathcal{F}_m)$ be an ordered m–tuple of flags in general position, where $\mathcal{F}_i := (V_i, \eta_i)$. Then the ordered 4–tuple of points $(V_1, V_2, V_3, \eta_1 \cap \eta_2)$ is in general position and its stabiliser in $\mathrm{SL}(3,\mathbb{R})$ is trivial.

It follows that the stabiliser of $\mathfrak{F}$ in $\mathrm{SL}(3,\mathbb{R})$ is also trivial, hence the action is free. ■

3.2 Triple ratio

Suppose $\mathfrak{F} = (\!(\mathcal{F}_0, \mathcal{F}_1, \mathcal{F}_2)\!)$ is a cyclically ordered triple of flags in general position. We define the *triple ratio* of $\mathfrak{F}$ as

$$\mathfrak{z}(\mathfrak{F}) := \frac{\overline{\eta_0}(\overline{V_1}) \cdot \overline{\eta_1}(\overline{V_2}) \cdot \overline{\eta_2}(\overline{V_0})}{\overline{\eta_0}(\overline{V_2}) \cdot \overline{\eta_1}(\overline{V_0}) \cdot \overline{\eta_2}(\overline{V_1})},$$

where $\overline{V_i} \in \mathbb{R}^3$ and $\overline{\eta_i} \in (\mathbb{R}^3)^*$ are representatives of V_i and η_i respectively. This is well-defined, as it is independent of the choice of representatives, and manifestly invariant under a cyclic permutation of the flags. The following property is easy to verify.

Lemma 3.2.1. *The triple ratio is invariant under projective transformations, that is:*

$$\mathfrak{z}(T \cdot \mathfrak{F}) = \mathfrak{z}(\mathfrak{F}) \qquad \textit{for all } \ T \in \mathrm{SL}(3,\mathbb{R}).$$

3.3 Cross ratio

Let $l \subset \mathbb{RP}^2$ be a line. Given $P_0, P_1, P_2, P_3 \in l$ with P_0, P_1, P_2 pairwise distinct, let $T : l \to \mathbb{RP}^1$ be the unique projective map such that $T(P_0) = \infty$, $T(P_1) = -1$ and $T(P_2) = 0$. Then the *cross ratio* of the ordered quadruple (P_0, P_1, P_2, P_3) is

$$\mathrm{CR}(P_0, P_1, P_2, P_3) = T(P_3).$$

We remark that this is *not* the standard definition of the cross ratio (as used, for example, in the definition of the Hilbert metric in §2.3), but an ingenious choice by Fock and Goncharov that results in their elegant coordinate system. Just as for the triple ratio, it is easy to check that the cross ratio is projectively invariant. Moreover, if x_0, x_1, x_2, x_3 are local coordinates for $P_0, P_1, P_2, P_3 \in l$, then

$$\mathrm{CR}(P_0, P_1, P_2, P_3) = \frac{(x_0 - x_1)(x_2 - x_3)}{(x_0 - x_3)(x_1 - x_2)}.$$

It follows that, if σ is a permutation on four symbols and $\lambda = \mathrm{CR}(P_0, P_1, P_2, P_3)$, then

$$\mathrm{CR}(P_{\sigma(0)}, P_{\sigma(1)}, P_{\sigma(2)}, P_{\sigma(3)}) \in \left\{\lambda, \frac{1}{\lambda}, 1 - \lambda, \frac{1}{1-\lambda}, 1 - \frac{1}{\lambda}, \frac{\lambda}{\lambda - 1}\right\}.$$

In particular $\mathrm{CR}(P_3, P_2, P_1, P_0) = \mathrm{CR}(P_0, P_1, P_2, P_3)$.

Similarly, we can define the cross ratio of four incident lines via duality. Let $P \in \mathbb{RP}^2$ be a point and l_0, l_1, l_2, l_3 lines through P with l_0, l_1, l_2 pairwise distinct.

Then the *cross ratio* of the ordered quadruple (l_0, l_1, l_2, l_3) is

$$\mathrm{CR}(l_0, l_1, l_2, l_3) := \mathrm{CR}(l_0^\perp, l_1^\perp, l_2^\perp, l_3^\perp).$$

Lemma 3.3.1. *Let $P \in \mathbb{RP}^2$ be a point and l_0, l_1, l_2, l_3 lines through P with l_0, l_1, l_2 pairwise distinct. Let m be a line intersecting the lines l_0, l_1, l_2, l_3 transversely in Q_0, Q_1, Q_2, Q_3 respectively, then $\mathrm{CR}(l_0, l_1, l_2, l_3) = \mathrm{CR}(Q_0, Q_1, Q_2, Q_3)$.*

Proof. Without loss of generality, we may assume that

$$P = \begin{bmatrix} 1 \\ 0 \\ 0 \end{bmatrix}, \quad Q_1 = \begin{bmatrix} 0 \\ 1 \\ -1 \end{bmatrix}, \quad Q_2 = \begin{bmatrix} 0 \\ 0 \\ 1 \end{bmatrix}.$$

It follows that

$$m = \big[1 : 0 : 0\big], \quad l_1 = \big[0 : 1 : 1\big], \quad l_2 = \big[0 : 1 : 0\big].$$

Since Q_0, Q_1, Q_2, Q_3 are collinear and Q_0, Q_1, Q_2 are pairwise distinct, there are numbers $a, b, c \in \mathbb{R}$ such that

$$Q_0 = \begin{bmatrix} 0 \\ 1 \\ a \end{bmatrix}, \quad Q_3 = \begin{bmatrix} 0 \\ b \\ c \end{bmatrix}, \quad \text{and} \quad l_0 = \big[0 : -a : 1\big], \quad l_3 = \big[0 : -c : b\big].$$

The projective transformation

$$M = \begin{bmatrix} 1 & 0 & 0 \\ 0 & 0 & -1 \\ 0 & 1 & 0 \end{bmatrix}$$

maps $(l_0^\perp, l_1^\perp, l_2^\perp, l_3^\perp)$ to (Q_0, Q_1, Q_2, Q_3), therefore

$$\mathrm{CR}(l_0, l_1, l_2, l_3) = \mathrm{CR}(Q_0, Q_1, Q_2, Q_3).$$ ■

The convention for the cross ratio used here was suggested by Fock and Goncharov [15, pg. 253]. It is motivated by the fact that, if P_0, P_1, P_2 are pairwise distinct points on a line $l \subset \mathbb{RP}^2$, we would like $\mathrm{CR}(P_0, P_1, P_2, P_3) > 0$ when P_3 and P_1 lie in different components of $l \backslash \{P_0, P_2\}$. Lemma 3.3.1 implies that if l_0, l_1, l_2 are pairwise distinct lines passing through a point $P \in \mathbb{RP}^2$, then $\mathrm{CR}(l_0, l_1, l_2, l_3) > 0$ when l_3 and l_1 lie in different regions of $\mathbb{RP}^2 \backslash \{l_0, l_2\}$. See Figure 3.1.

Henceforth, if $P, Q \in \mathbb{RP}^2$ are two points and $m, l \subset \mathbb{RP}^2$ are two lines, we will denote by PQ the line passing through P and Q, and by lm the point of intersection between l and m.

Theorem 3.3.2. *Let $\mathfrak{F} = (\!(\mathcal{F}_0, \mathcal{F}_1, \mathcal{F}_2)\!)$ be a cyclically ordered triple of flags in*

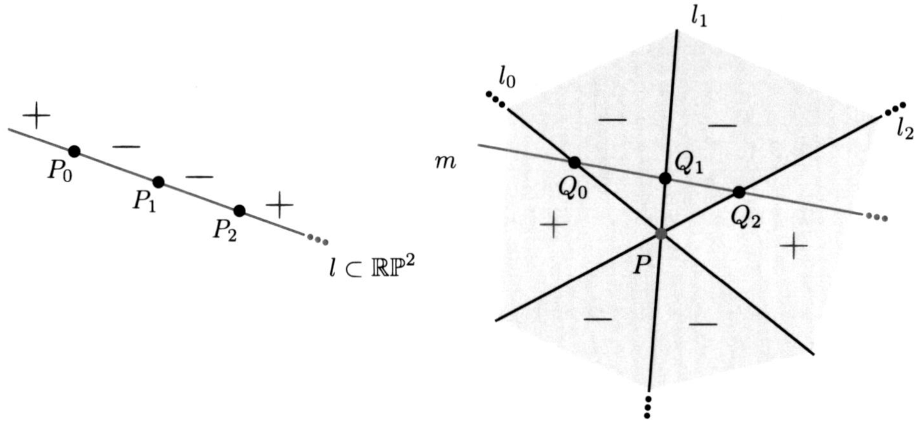

Figure 3.1: The points P_3 and P_1 lie in different components of $l \backslash \{P_0, P_2\}$ if and only if $\mathrm{CR}(P_0, P_1, P_2, P_3) > 0$. Likewise, l_3 and l_1 lie in different regions of $\mathbb{RP}^2 \backslash \{l_0, l_2\}$ when $\mathrm{CR}(l_0, l_1, l_2, l_3) > 0$.

general position where $\mathcal{F}_i = (V_i, \eta_i)$. *Then*

$$\mathrm{CR}(\eta_0, V_0V_1, V_0(\eta_1\eta_2), V_0V_2) = \mathfrak{Z}(\mathfrak{F}).$$

Proof. Both cross ratio and triple ratio are projectively invariant, and the points $V_0, V_1, V_2, \eta_1\eta_2$ are in general position so we may assume, without loss of generality, that

$$V_0 = \begin{bmatrix} 0 \\ 0 \\ 1 \end{bmatrix}, \quad V_1 = \begin{bmatrix} 1 \\ 0 \\ 1 \end{bmatrix}, \quad V_2 = \begin{bmatrix} 0 \\ 1 \\ 1 \end{bmatrix}, \quad \eta_1\eta_2 = \begin{bmatrix} 1 \\ 1 \\ 1 \end{bmatrix}.$$

It follows that

$$\eta_1 = \begin{bmatrix} 1 : 0 : -1 \end{bmatrix}, \quad \eta_2 = \begin{bmatrix} 0 : 1 : -1 \end{bmatrix}.$$

The projective line η_0 passes through V_0 but does not pass through V_2 or V_1, so $\eta_0 = [A : 1 : 0]$ for some $A \in \mathbb{R} \backslash \{0\}$. By Lemma 3.3.1,

$$\mathrm{CR}(\eta_0, V_0V_1, V_0(\eta_1\eta_2), V_0V_2) = \mathrm{CR}(\eta_0\eta_2, (V_0V_1)\eta_2, (V_0(\eta_1\eta_2))\eta_2, (V_0V_2)\eta_2).$$

A direct calculation shows that

$$\mathfrak{Z}(\mathfrak{F}) = A = \mathrm{CR}(\eta_0\eta_2, (V_0V_1)\eta_2, (V_0(\eta_1\eta_2))\eta_2, (V_0V_2)\eta_2).$$

This completes the proof. ∎

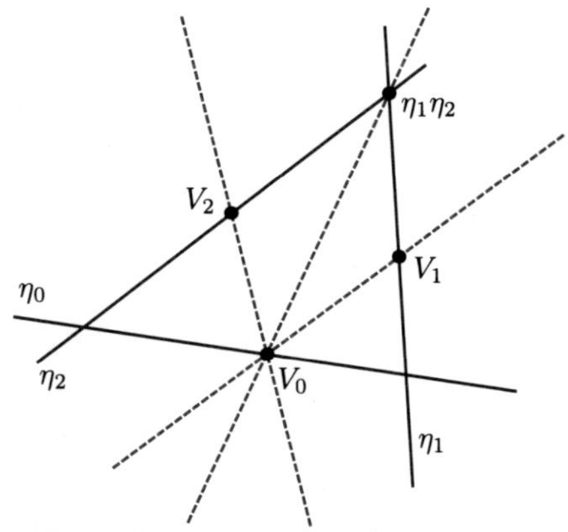

Figure 3.2: The cross ratio $\mathrm{CR}(\eta_0, V_0V_1, V_0(\eta_1\eta_2), V_0V_2)$ is equal to the triple ratio $\not{3}\,(\!(\mathcal{F}_0, \mathcal{F}_1, \mathcal{F}_2)\!)$.

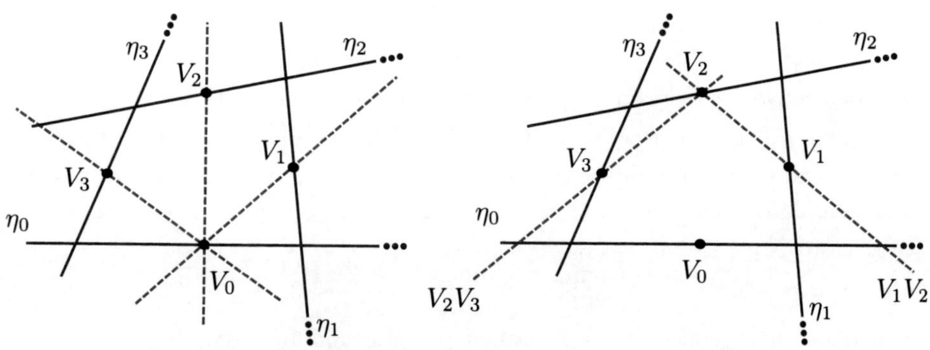

Figure 3.3: The quadruple ratio $\not{4}(\mathfrak{F})$ can be read off both in terms of $\mathrm{CR}(\eta_0, V_0V_3, V_0V_2, V_0V_1)$ and $\not{3}\,(\!((V_0, \eta_0), (V_3, V_2V_3), (V_1, V_1V_2))\!)$.

3.4 Quadruple ratio

Let $\mathfrak{F} = (\mathcal{F}_0, \mathcal{F}_1, \mathcal{F}_2, \mathcal{F}_3)$ be an ordered quadruple of flags in general position, with $\mathcal{F}_i = (V_i, \eta_i)$. We define the *quadruple ratio* of $\mathfrak{F}$ as

$$\not{4}(\mathfrak{F}) := \mathrm{CR}(\eta_0, V_0V_3, V_0V_2, V_0V_1).$$

The ratio $\not{4}(\mathfrak{F})$ is sometimes referred as *edge ratio* with respect to V_0V_2. We give an intuitive description of this definition in §3.7. It follows from Theorem 3.3.2 that

$$\not{4}(\mathfrak{F}) = \not{3}\,(\!((V_0, \eta_0), (V_3, V_2V_3), (V_1, V_1V_2))\!)\ .$$

The requirement that $(\mathcal{F}_0, \mathcal{F}_1, \mathcal{F}_2, \mathcal{F}_3)$ are flags in general position may be relaxed, but this is not needed in these notes.

3.5 Pairs of convex polygons and configuration of flags

We are now interested in the situation, where m flags form two convex polygons, one inscribed into the other. We parametrise such configurations for $m = 3, 4$ using the triple ratio and quadruple ratio. This will be a key ingredient in the proof of Theorem 4.4.1.

We say that a non-degenerate m–gon σ_1 in an affine patch is *strictly inscribed* into another non-degenerate m–gon σ_2 in that affine patch if each edge of σ_2 contains one vertex of σ_1 in its interior. We orient the affine patch. An m–gon is topologically a circle, therefore it may be oriented in two ways. A pair of oriented m–gons is *coherently oriented* if the two m–gons have the same orientation. Furthermore, a coherently oriented pair of strictly inscribed m–gons is *marked* by a choice of a preferred vertex of the inscribed polygon (equiv. a preferred edge of the circumscribed polygon) and subsequent ordering of the vertices of the inscribed polygon (equiv. the edges of the circumscribed polygon) starting from the preferred vertex and following the orientation of the pair.

Let $\mathcal{P}_m$ be the space of coherently oriented pairs of strictly inscribed convex m–gons in $\mathbb{RP}^2$, and $\mathcal{P}_m^*$ be the space of marked coherently oriented pairs of strictly inscribed convex m–gons in $\mathbb{RP}^2$, both modulo the action of $\mathrm{SL}(3,\mathbb{R})$.

Similarly, Conf_m is the space of cyclically ordered m–tuples of flags in general position and Conf_m^* is the space of ordered m–tuples of flags in general position, both modulo the action of $\mathrm{SL}(3,\mathbb{R})$. There are natural maps

$$\xi_m : \mathcal{P}_m \to \mathrm{Conf}_m, \qquad \text{and} \qquad \xi_m^* : \mathcal{P}_m^* \to \mathrm{Conf}_m^*,$$

defined by the vertices of the inscribed polygon and by extending the edges of the circumscribed polygon to lines. The fact that the polygons are assumed to be strictly inscribed one into the other ensures that the corresponding flags are in general position. Moreover, two pairs of polygons are projectively equivalent if and only if the associated flags are projectively equivalent. Hence ξ_m and ξ_m^* are injective and hence $\mathcal{P}_m$ and $\mathcal{P}_m^*$ can be identified with their images in Conf_m and Conf_m^* respectively. On the other hand, these maps are not surjective (see §3.6 below). We give a characterisation of $\mathcal{P}_3$ and $\mathcal{P}_4^*$ in §3.7.

3.6 Inscribed triangles and triples of flags

We present an example where we relate $\mathcal{P}_3$, Conf_3 and convex domains in $\mathbb{RP}^2$. Let

$$v_A := \begin{bmatrix} 1 \\ 0 \\ 0 \end{bmatrix}, \qquad v_B := \begin{bmatrix} 0 \\ 0 \\ 1 \end{bmatrix}, \qquad v_C := \begin{bmatrix} 1 \\ 1 \\ 1 \end{bmatrix},$$

and

$$\eta_C := [-t : 1 + t : -1], \qquad \text{for } t \in \mathbb{R}.$$

Now consider the cyclically ordered triple of flags $\mathfrak{F} = (\!(\mathcal{F}_A, F_B, \mathcal{F}_C)\!)$, where

$$\mathcal{F}_A := (v_A, v_B^\perp), \qquad \mathcal{F}_B := (v_B, v_A^\perp), \qquad \mathcal{F}_C := (v_C, \eta_C).$$

This triple is in general position if and only if $t \neq 0, -1$, as these are the cases where the line η_C passes through the points v_A and $v_A^\perp v_B^\perp$ respectively. The case $t = \infty$ would also violate generality as this would ensure that η_C passes through v_B.

There are four triangular regions in $\mathbb{RP}^2$ with vertices v_A, v_B and v_C. The triple of flags $\mathfrak{F}$ is in the image of ξ_3 if and only if one of these triangles is disjoint from each of the lines $v_B^\perp$, $v_A^\perp$ and η_C. The lines $v_B^\perp$ and $v_A^\perp$ pass through three of the four potential triangles. The only remaining option is the triangle which is strictly contained in the affine patch $\mathcal{A} := \{[x : y : 1]^\intercal \mid x, y \in \mathbb{R}\} \subset \mathbb{RP}^2$. Therefore, $\mathfrak{F} \in \xi_3(\mathcal{P}_3)$ if and only if η_C intersects the line $v_A v_B$ at a point of the form $[x : 0 : 1]^\intercal$, where $x < 0$. This is the case if and only if $t > 0$. The affine patch $\mathcal{A}$ is depicted in Figure 3.4.

A direct calculation shows that $\mathfrak{Z}(\mathfrak{F}) = t$. So the triple ratio may be used to determine whether $\mathfrak{F} \in \xi_3(\mathcal{P}_3)$. Moreover, there is a unique conic $\mathcal{C}$ passing through v_A, v_B and v_C with supporting projective lines $v_A^\perp$ and $v_B^\perp$, namely the set of points

$$\begin{bmatrix} x^2 \\ xz \\ z^2 \end{bmatrix}, \qquad \text{for } (x, z) \in \mathbb{R}^2 \backslash (0, 0).$$

The line defined by η_C is a supporting line to $\mathcal{C}$ at v_C if and only if $t = 1$. As projective transformations preserve the set of conics, this triple ratio test may be used to determine whether an element of $\mathcal{P}_3$ inscribes a conic. We will expand further upon this in §6.2.

3.7 Parametrisation of the spaces $\mathcal{P}_3$ and $\mathcal{P}_4^*$

We give a parametrisation of $\mathcal{P}_3$ and $\mathcal{P}_4^*$ using triple ratios and quadruple ratios. This will be used in the proof of Theorem 4.4.1. We define the map $\mathfrak{Z}\colon \mathcal{P}_3 \to \mathbb{RP}^1$ as follows. For $\chi \in \mathcal{P}_3$, choose a representative $\mathfrak{F}$ of $\xi_3(\chi)$. Then $\mathfrak{Z}(\chi) := \mathfrak{Z}(\xi_3(\chi)) = \mathfrak{Z}(\mathfrak{F})$ is well-defined by Lemma 3.2.1.

Theorem 3.7.1. *The map $\mathfrak{Z}$ establishes a bijection between $\mathcal{P}_3$ and $\mathbb{R}_{>0}$.*

Proof. Let $\chi \in \mathcal{P}_3$ and fix a representative $\mathfrak{F} = (\!(\mathcal{F}_0, \mathcal{F}_1, \mathcal{F}_2)\!)$ of $\xi_3(\chi)$, where $\mathcal{F}_i = (V_i, \eta_i)$ for $i = 0, 1, 2$. As in Theorem 3.3.2, after a projective transformation, we can assume

$$V_0 = \begin{bmatrix} 0 \\ 0 \\ 1 \end{bmatrix}, \quad V_1 = \begin{bmatrix} 1 \\ 0 \\ 1 \end{bmatrix}, \quad V_2 = \begin{bmatrix} 0 \\ 1 \\ 1 \end{bmatrix}, \quad \eta_1\eta_2 = \begin{bmatrix} 1 \\ 1 \\ 1 \end{bmatrix},$$

$$\eta_1 = \left[1 : 0 : -1\right], \quad \eta_2 = \left[0 : 1 : -1\right], \qquad \eta_0 = \left[t : 1 : 0\right], \quad t \in \mathbb{R}.$$

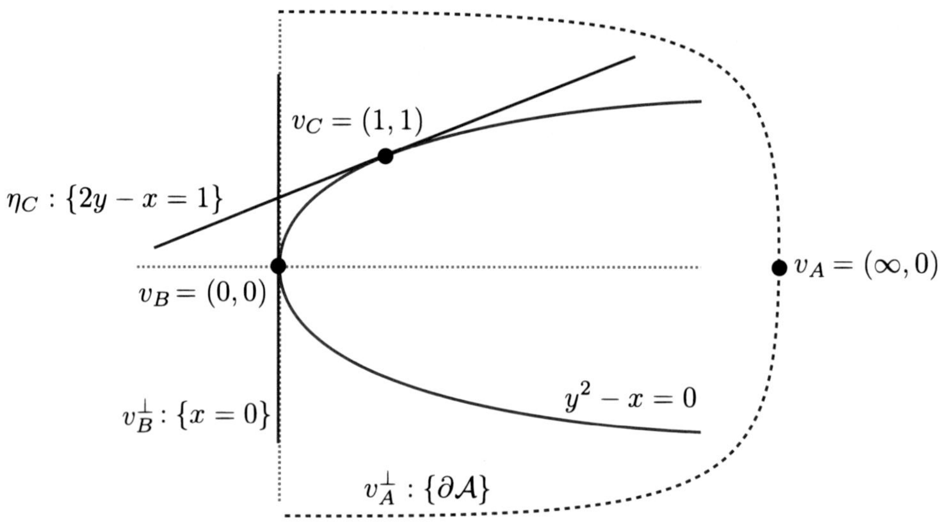

Figure 3.4: When $t = 1$, the cyclically ordered triple $(\!(\mathcal{F}_A, \mathcal{F}_B, \mathcal{F}_C)\!)$ corresponds to an element of $\mathcal{P}_3$ and the line $\eta_C = [-1:2:-1]$ is a supporting projective line to $\mathcal{C}$ at the point v_C.

Recall that $t = \mathfrak{Z}(\mathfrak{F}) \in \mathbb{R}\backslash\{0\}$ because η_0 is a line through V_0 which is disjoint from V_2 and V_1. In this setting, χ is uniquely determined by $\mathfrak{Z}(\mathfrak{F})$ so injectivity is immediate. Theorem 3.3.2 and the discussion in §3.3 imply that $\mathfrak{Z}(\mathfrak{F})$ is positive if and only if η_0 and $V_0(\eta_1\eta_2)$ lie in different regions of $\mathbb{RP}^2\backslash\{V_0V_1, V_0V_2\}$. It follows that $\mathfrak{Z}(\mathfrak{F}) > 0$ if and only if $\mathfrak{F}$ represents an element of $\mathcal{P}_3$, proving that $\mathfrak{Z}(\mathcal{P}_3) \subseteq \mathbb{R}_{>0}$. Surjectivity is straightforward, and follows from the discussion in § 3.6. ■

Now we are going to show something similar for $\mathcal{P}_4^*$ using both triple ratios and quadruple ratios. We define $g\colon \mathcal{P}_4^* \to (\mathbb{RP}^1)^4$ as follows. For $\chi \in \mathcal{P}_4^*$, choose a representative $\mathfrak{F} = (\mathcal{F}_0, \mathcal{F}_1, \mathcal{F}_2, \mathcal{F}_3)$ of $\xi_4^*(\chi)$, with $\mathcal{F}_i = (V_i, \eta_i)$ for $i = 0,1,2,3$. Then $g(\chi) := (t_{012}, t_{023}, e_{02}, e_{20})$, where

$$\begin{aligned}
t_{012} &:= \mathfrak{Z}\,(\!(\mathcal{F}_0, \mathcal{F}_1, \mathcal{F}_2)\!),\\
t_{023} &:= \mathfrak{Z}\,(\!(\mathcal{F}_0, \mathcal{F}_2, \mathcal{F}_3)\!),\\
e_{02} &:= \mathfrak{A}(\mathcal{F}_0, \mathcal{F}_1, \mathcal{F}_2, \mathcal{F}_3) = \mathfrak{Z}\,(\!((V_0, \eta_0), (V_3, V_2V_3), (V_1, V_1V_2))\!),\\
e_{20} &:= \mathfrak{A}(\mathcal{F}_2, \mathcal{F}_3, \mathcal{F}_0, \mathcal{F}_1) = \mathfrak{Z}\,(\!((V_2, \eta_2), (V_1, V_0V_1), (V_3, V_3V_0))\!)\,.
\end{aligned}$$

Once again, g is well-defined due to the projective invariance of triple ratio. One may visualise χ as in Figure 3.5, with an additional edge crossing from V_0 to V_2. This motivates why $\mathfrak{A}(\mathfrak{F})$ is also called the *edge ratio* of the shown segment of the line V_0V_2 oriented from V_0 to V_2. We remark that the edge ratio as defined here is the reciprocal of the edge ratio defined in [15].

Theorem 3.7.2. *The map g establishes a bijection between $\mathcal{P}_4^*$ and $\mathbb{R}^4_{>0}$.*

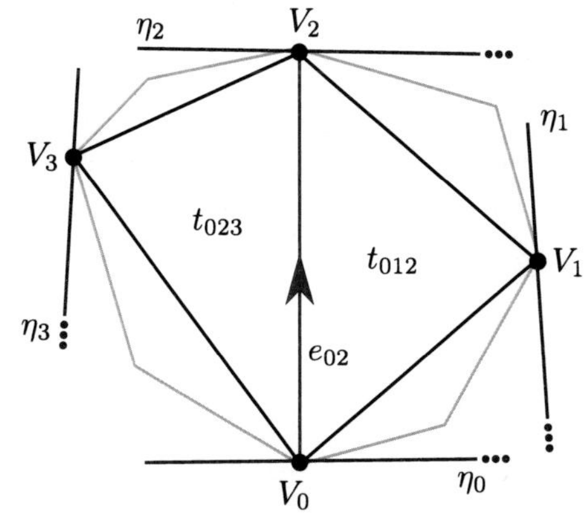

Figure 3.5: The edge ratio e_{02} is associated to the shown segment of the line V_0V_2 oriented from V_0 to V_2, adjacent to the two triangles with triple ratios t_{012}, t_{023}.

Proof. Let $\chi \in \mathcal{P}_4^*$ and choose a representative $\mathfrak{F} = (\mathcal{F}_0, \mathcal{F}_1, \mathcal{F}_2, \mathcal{F}_3)$ of $\xi_4^*(\chi)$, with $\mathcal{F}_i = (V_i, \eta_i)$. We start with few observations about the relation between the pair of quadrilaterals χ and the ratios $g(\chi)$.

As in the proof of Theorem 3.7.1, we assume without loss of generality that V_0, V_2, V_3 and $\eta_0\eta_2$ are fixed at an arbitrary generic quadruple of points in some affine patch $\mathcal{A}$ of $\mathbb{RP}^2$.

Having fixed these points, the lines η_2 and η_0 are determined. By Theorem 3.7.1, the line η_3 is uniquely determined by t_{023}, and any assignment of positive real number to t_{023} gives rise to a unique, well-defined pair of triangles of $\mathcal{P}_3$.

If necessary, we rearrange the affine patch $\mathcal{A}$ to contain this pair of triangles. In $\mathcal{A}$, a polygon may be referred to as the convex hull of its vertices without ambiguity.

Recall from §3.4 that the quadruple ratios e_{02} and e_{20} can be expressed as triple ratios

$$e_{02} = \mathfrak{Z}\,(((V_0, \eta_0), (V_3, V_2V_3), (V_1, V_1V_2))),$$

and

$$e_{20} = \mathfrak{Z}\,(((V_2, \eta_2), (V_1, V_1V_0), (V_3, V_3V_0))),$$

or equivalently as cross ratios

$$e_{02} = \mathrm{CR}(\eta_0, V_0V_3, V_0V_2, V_0V_1) \quad \text{and} \quad e_{20} = \mathrm{CR}(\eta_2, V_2V_1, V_2V_0, V_2V_3).$$

As in Theorem 3.7.1, the lines V_0V_1 and V_2V_1 are uniquely determined by e_{02} and e_{20}, so V_1 is uniquely determined. Furthermore, V_1 belongs to the triangle $\langle V_0, V_2, \eta_0\eta_2\rangle$ if and only if $e_{02} > 0$ and $e_{20} > 0$. Equivalently, the quadrilateral $\langle V_0, V_1, V_2, V_3\rangle$ in $\mathcal{A}$ is convex and inscribed in the triangle bounded by the lines η_0, η_2 and η_3 if and only if $e_{02}, e_{20} \in \mathbb{R}_{>0}$. This is a necessary and sufficient condition for $\langle V_0, V_1, V_2, V_3\rangle$ to be the inscribed quadrilateral in $\chi \in \mathcal{P}_4^*$. This construction is

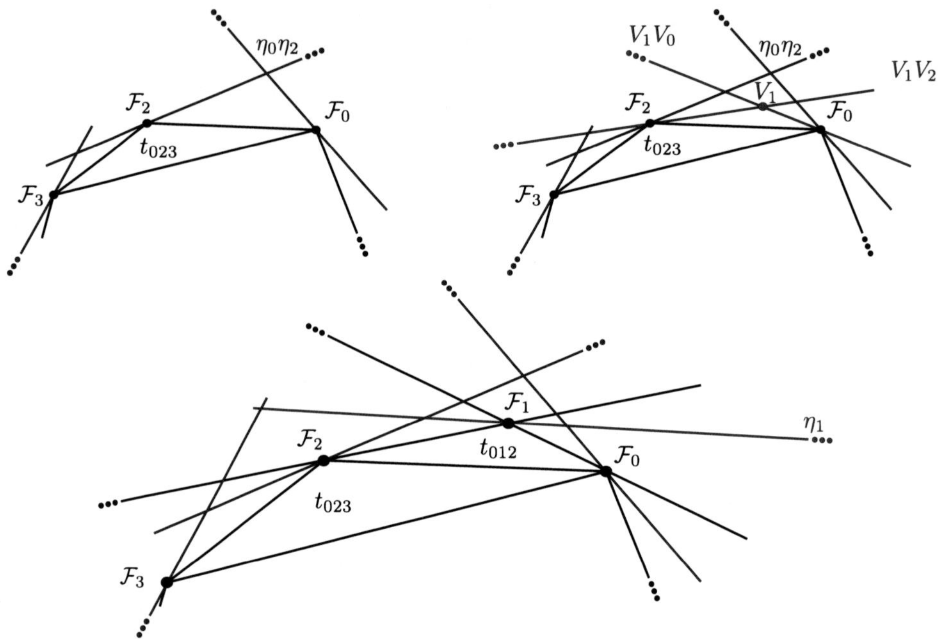

Figure 3.6: The lines V_1V_0 and V_1V_2 are uniquely determined by the ratios $\not 4(\mathcal{F}_0, \mathcal{F}_1, \mathcal{F}_2, \mathcal{F}_3)$ and $\not 4(\mathcal{F}_2, \mathcal{F}_3, \mathcal{F}_0, \mathcal{F}_1)$, while η_1 is uniquely determined by V_1 and $\not 3\,(\!(\mathcal{F}_0, \mathcal{F}_1, \mathcal{F}_2)\!)$.

shown in Figure 3.6.

Given that V_1, $\mathcal{F}_0$ and $\mathcal{F}_2$ are now fixed, the line η_1 is uniquely determined by t_{012}, once again appealing to Theorem 3.7.1. Since χ belongs to $\mathcal{P}_4^*$, η_1 does not intersect the triangle $\langle V_0, V_1, V_2\rangle$. That happens if and only if $t_{012} > 0$.

The fact that χ is uniquely determined by the four numbers

$$g(\chi) = (t_{012}, t_{023}, e_{02}, e_{20})$$

implies that g is injective. The above discussion also shows that $g(\mathcal{P}_4^*) \subseteq \mathbb{R}^4_{>0}$. Surjectivity of g follows from the surjectivity of $\not 3$ and Theorem 3.3.2. ■

The *cyclic group of order four* $C_4 := \langle \alpha \mid \alpha^4 = 1\rangle$, acts on $\mathcal{P}_4^*$ by changing the marking, namely $\mathcal{P}_4^*/C_4 \cong \mathcal{P}_4$. The corresponding action of C_4 on $\mathbb{R}^4_{>0}$ can be thought of as the change of coordinates:

$$\alpha \cdot (t_{012}, t_{023}, e_{02}, e_{20}) := \left(t'_{123}, t'_{013}, e'_{13}, e'_{31}\right),$$

where

$$t'_{123} = \frac{t_{012}(e_{02}t_{023}e_{20} + e_{02}t_{023} + e_{02} + 1)}{e_{02}t_{012}e_{20} + t_{012}e_{20} + e_{20} + 1}, \qquad e'_{13} = \frac{e_{20} + 1}{(e_{02} + 1)t_{012}e_{20}},$$

$$t'_{013} = \frac{t_{023}(e_{02}t_{012}e_{20} + t_{012}e_{20} + e_{20} + 1)}{e_{02}t_{023}e_{20} + e_{02}t_{023} + e_{02} + 1}, \qquad e'_{31} = \frac{e_{02} + 1}{e_{02}t_{023}(e_{20} + 1)}.$$

This change of coordinates will be analysed again in §5.1.

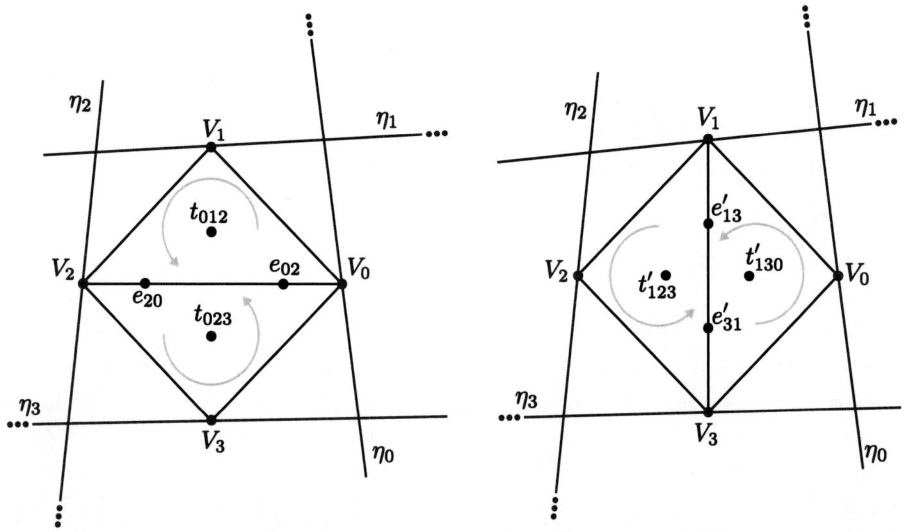

Figure 3.7: The change of coordinates for the action of C_4. The quadrilateral formed by V_i, $i = 0, \ldots, 3$ is depicted in an affine patch such that the triangles are oriented anticlockwise.

Chapter 4

The Parametrisation of the Moduli Space

Ratios of flags can be viewed as the algebraic underpinning of Fock and Goncharov's parametrisation of the moduli spaces of interest in these notes. In addition, the parametrisation is based in geometry and topology through framings of the ends and ideal triangulations. We will now describe these ingredients and then give a self-contained construction of the Fock-Goncharov parametrisation. Our proofs in this section are different from those found in [15].

4.1 Framing the ends

We described the spaces $\mathcal{T}_3^+(S)$ and $\mathcal{T}_3^-(S)$ corresponding to framed structures where each end is either *minimal* or *maximal*, and the distinction between the spaces arises from the framings that are allowed for a hyperbolic end. In this section, we give a different definition of these spaces and show in the proofs of Theorems 4.4.1 and 4.7.1 that these definitions are in fact equivalent. We refer the reader to the classification of ends as hyperbolic, special and cusps, as well as the definition of maximal and minimal hyperbolic ends given in §2.2.8.

Each end may be endowed with a *framing.* There are two definitions which are related by projective duality. A *positive framing* of $(\Omega, \Gamma, f) \in \mathcal{T}_3^m(S)$ is a choice, for each end $\mathcal{E}_k$, of a pair consisting of:

- If $\mathcal{E}_k$ is a maximal hyperbolic end: the Γ–orbit of the saddle point V of Γ_k and the Γ–orbit of a Γ_k–invariant supporting line η through V.
- Otherwise: the Γ–orbit of a fixed point V of Γ_k in the frontier of Ω, and the Γ–orbit of a Γ_k–invariant supporting line η through V.

Under this definition, there are two possible positive framings for each maximal hyperbolic end, for each minimal hyperbolic end there are four, and for each special end there are three. Each cusp has a unique positive framing. We recall that $\mathcal{T}_3^\times(S)$ is the space of framed structures on S. Hence two positively framed structures are equivalent if and only if they represent the same element of $\mathcal{T}_3^\times(S)$.

We introduce the following notation. We denote by $\mathcal{T}_3^+(S)$ the set of equivalence classes of *positively framed marked properly convex projective structures with maximal or minimal hyperbolic ends* on S.

We call π^+ the natural projection of $\mathcal{T}_3^+(S)$ onto $\mathcal{T}_3^m(S)$ defined by forgetting the framing. A structure (Ω, Γ, f) with a fixed positive framing will be denoted by $(\Omega, \Gamma, f)^+$.

A *negative framing* of $(\Omega, \Gamma, f) \in \mathcal{T}_3^m(S)$ is a choice, for each end $\mathcal{E}_k$, of a pair consisting of:

- If $\mathcal{E}_k$ is a minimal hyperbolic end: the Γ–orbit of the supporting line η, that is disjoint from the saddle point and invariant under Γ_k, and the Γ–orbit of a fixed point V of Γ_k on η.
- Otherwise: the Γ–orbit of a fixed point V of Γ_k in the frontier of Ω, and the Γ–orbit of a Γ_k–invariant supporting line η through V.

A *negative framing* differs from a positive framing only at hyperbolic ends. For a positive framing, the choice of the pair consisting of a non-saddle point and the line through it and the saddle point was only allowed for minimal hyperbolic ends. In a *negative framing*, this choice is only allowed for maximal hyperbolic ends instead, while keeping everything else the same. This means that for each maximal hyperbolic end there are four possible choices, while for each minimal hyperbolic end there are now two. We denote the structure (Ω, Γ, f) with a fixed negative framing by $(\Omega, \Gamma, f)^-$. The set $\mathcal{T}_3^-(S)$ is the set of equivalence classes of *negatively framed marked properly convex projective structures with maximal or minimal hyperbolic ends* on S, and $\pi^- : \mathcal{T}_3^-(S) \to \mathcal{T}_3^m(S)$ is the associated projection map.

The reason for distinguishing between positively and negatively framed properly convex projective structures will be elucidated in the proof of Theorem 4.4.1 when we construct framed properly convex projective structures using a finite set of numerical data and we need to make a choice of whether to use the positive or negative framing convention. We will subsequently find that projective duality induces natural *duality maps* $\sigma^+ : \mathcal{T}_3^+(S) \to \mathcal{T}_3^-(S)$ and $\sigma^- : \mathcal{T}_3^-(S) \to \mathcal{T}_3^+(S)$ which are mutually inverse. This will be discussed in detail in §4.6.

It will be shown in §6.1 that finite-volume structures on S are those in which every end is a cusp. In particular, finite-volume structures have a unique framing so there are injective maps $\mathcal{T}_3^f(S) \hookrightarrow \mathcal{T}_3^+(S)$ and $\mathcal{T}_3^f(S) \hookrightarrow \mathcal{T}_3^-(S)$.

4.2 On horospheres at cusps and boundary components

It was already mentioned in the introduction that Fock and Goncharov offer an interpretation of cusps and geodesic boundary components depending on the flags used to decorate the ends; the different cases can be glimpsed from Figure 2.6. In this section, we offer a different interpretation, using *algebraic horospheres*, which results in a different classification for special ends.

Let us first recall some definitions (see [10, Section 3 pg. 201] for details). Let Ω be a properly convex domain in $\mathbb{RP}^2$, $p \in \partial\overline{\Omega}$ and $H \subset \mathbb{RP}^2$ a supporting hyperplane to Ω at p. Then $\mathrm{SL}(H,p)$ is the subgroup of $\mathrm{SL}(3,\mathbb{R})$ that preserves both H and p. Let $\mathcal{G} = \mathcal{G}(H,p)$ be the subgroup of $\mathrm{SL}(H,p)$ consisting of those elements $A \in \mathrm{SL}(H,p)$, which satisfy:

- A acts as the identity on H,
- $A(\eta) = \eta$ for every line η in $\mathbb{RP}^2$ passing through p.

If η is a line containing p that is not contained in H, then $\mathcal{G}$ acts on η as the group $\mathrm{Par}(\eta, p)$ of parabolic transformations of η fixing p, thus there are isomorphisms $(\mathbb{R}, +) \cong \mathrm{Par}(\eta, p) \cong \mathcal{G}$.

Let $\mathcal{S}_0 \subset \partial\overline{\Omega}$ be the subset of $\partial\overline{\Omega}$ obtained by deleting p and all line segments in $\partial\overline{\Omega}$ with one endpoint at p. A *generalized horosphere centred on* (H, p) is the image $\mathcal{S}_A := A(\mathcal{S}_0)$ of $\mathcal{S}_0$ under $A \in \mathcal{G}(H, p)$. An *algebraic horosphere*, or simply *horosphere*, is a generalized horosphere contained in Ω.

If $B \in \mathrm{SL}(H, p)$ preserves Ω, it is well-known that B acts on horospheres by

$$B(\mathcal{S}_A) = BA(\mathcal{S}_0) = BAB^{-1}(B\mathcal{S}_0) = BAB^{-1}(\mathcal{S}_0) = \mathcal{S}_{BAB^{-1}}.$$

This action can be expressed in terms of the *exponential horosphere displacement function* τ, defined as follows. Given $B \in \mathrm{SL}(H, p)$, let $\lambda_+(B)$ be the eigenvalue for the eigenvector p. If $v \in \mathbb{RP}^2 \backslash H$, then $Bv + H = \lambda_- v + H$ and $\lambda_- = \lambda_-(B)$ is another eigenvalue of B. This does not depend on the choice of v. The *exponential horosphere displacement function* with respect to (H, p) is the homomorphism

$$\tau : \mathrm{SL}(H, p) \to (\mathbb{R}^\times, \times) \quad \text{given by} \quad \tau(B) = \frac{\lambda_+(B)}{\lambda_-(B)}.$$

Choose an isomorphism from $(\mathbb{R}, +)$ to $\mathrm{Par}(\eta, p)$ given by $t \mapsto A_t$ and denote

$$\mathcal{S}_t = A_t(\mathcal{S}_0).$$

Theorem 4.2.1 (Cooper-Long-Tillmann 2015, [10]). *If $B \in \mathrm{SL}(H, p)$ preserves Ω, then $B(\mathcal{S}_t) = \mathcal{S}_{\tau(B)t}$.*

Let $x \in \mathcal{T}_3^+(S)$, and let γ be a generator of the peripheral subgroup $\pi_1(\mathcal{E})$. Let $\Omega := \mathrm{dev}(\widetilde{S})$. If $\mathrm{hol}(\gamma)$ is totally hyperbolic, it fixes three pairwise distinct points $\{V_{\lambda_1}, V_{\lambda_2}, V_{\lambda_3}\}$ of $\mathbb{RP}^2$ and preserves the lines $\{H_{12}, H_{13}, H_{23}\}$ through them (Figure 2.6 on the left). We remind the reader that Ω may or may not contain the shaded triangle. Horospheres centred at (H_{ij}, V_{λ_i}) look different depending on the triple $(\Omega, V_{\lambda_i}, H_{ij})$, (the different horospheres are depicted in Figure 4.1). In all cases $\tau(B) \neq 1$, therefore $\mathrm{hol}(\gamma)$ does not preserve the horospheres.

When $\mathrm{hol}(\gamma)$ is parabolic, it fixes a unique point $V_{\lambda_1} \in \partial\overline{\Omega}$ and preserves a unique supporting hyperplane H through V_{λ_1}. Since $\tau(B) = 1$, Theorem 4.2.1 implies that $\mathrm{hol}(\gamma)$ preserves horospheres at (H, p), as depicted in Figure 4.2. Hence horospheres are $\mathrm{hol}(\gamma)$–orbits in Ω.

When $\mathrm{hol}(\gamma)$ is quasi-hyperbolic, we have a more subtle situation, in between the two previous ones. In this case $\mathrm{hol}(\gamma)$ fixes two points $\{V_{\lambda_1}, V_{\lambda_2}\}$ on $\partial\overline{\Omega}$, preserves the line H_{12} through them and acts as a unipotent transformation on a second line H_1 containing only V_{λ_1}. On one hand, the exponential horosphere displacements with respect to (H_{12}, V_{λ_2}) and (H_1, V_{λ_1}) are non-trivial, behaving like a hyperbolic end. Whereas for (H_{12}, V_{λ_1}), $\tau(\mathrm{hol}(\gamma)) = 1$ and algebraic horospheres are preserved, analogously to the parabolic case. We demonstrate this situation in Figure 4.3.

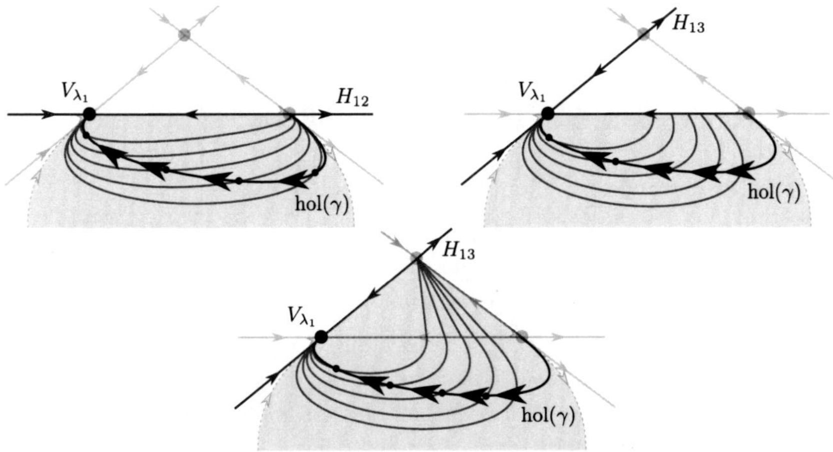

Figure 4.1: Horospheres are not preserved by a totally hyperbolic peripheral element.

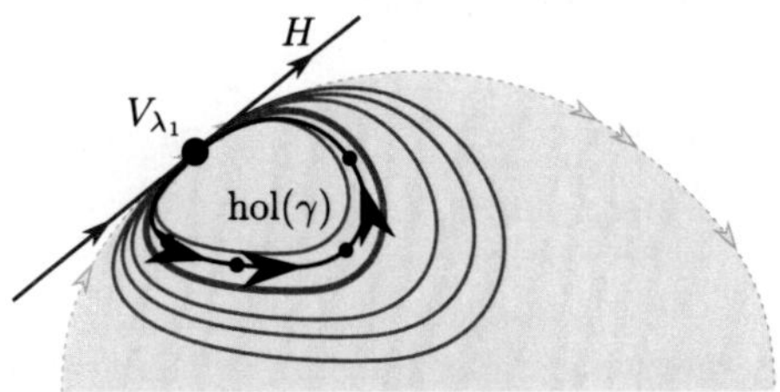

Figure 4.2: Horospheres are preserved at a cusp.

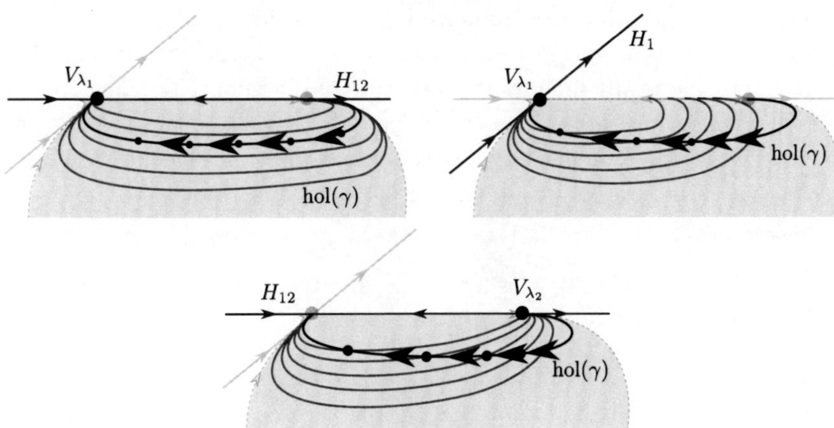

Figure 4.3: Special boundary components preserve some horospheres and permute others.

4.3 Ideal triangulations of surfaces

By construction, $S_{g,n} = S_g \backslash \mathcal{P}$, where $\mathcal{P} = \{p_1, \ldots, p_n\}$. An *essential arc* in S_g is an arc $\alpha \subset S_g$ whose endpoints are in $\mathcal{P}$ (they may coincide) and whose interior is embedded in $S_{g,n}$, with the condition that if $S_g \backslash \alpha$ has two components, then neither component is a disc meeting $\mathcal{P}$ only in the endpoints of α.

An *ideal triangulation* T of $S_{g,n}$ is the intersection with $S_{g,n}$ of a maximal collection of pairwise disjoint essential arcs in S_g, no two of which are ambient isotopic fixing $\mathcal{P}$. We regard two ideal triangulations of $S_{g,n}$ as equivalent if they are isotopic via an isotopy of S_g that fixes $\mathcal{P}$.

The essential arcs in an ideal triangulation cut $S_{g,n}$ into *ideal triangles.* Each component of $S_{g,n} \backslash \mathrm{T}$ is the interior of an *ideal triangle*, and the closure of this component in $S_{g,n}$ is an ideal triangle. We note that some ideal triangles may have two edges identified. Whence as a subsurface of $S_{g,n}$, an ideal triangle is either a closed disc with three punctures on its boundary, or a closed disc with one puncture on its boundary and one puncture in its interior.

The following lemma is a direct application of the fact that S_g admits a triangulation with a single vertex:

Lemma 4.3.1. *The surface $S_{g,n}$ admits an ideal triangulation. Moreover, every ideal triangulation has $-2\chi(S_{g,n})$ ideal triangles.*

An *edge flip* on an ideal triangulation consists of picking two distinct ideal triangles sharing an edge, removing that edge and replacing it with the other diagonal of the square thus formed.

For instance, any ideal triangulation of $S_{1,1}$ has three essential arcs and divides the surface into two ideal triangles. All of these ideal triangulations are combinatorially equivalent. However, performing an edge flip results in a non-isotopic ideal triangulation. The space of isotopy classes of ideal triangulations of the once-punctured torus naturally inherits the structure of the infinite trivalent tree, where vertices correspond to isotopy classes of ideal triangulations, and there is an edge between two such classes if and only if they are related by an edge flip. A well-known geometric realisation of this was described by Floyd and Hatcher [13].

In general, we have:

Lemma 4.3.2 (Hatcher [27], Mosher [40])**.** *Any two ideal triangulations of S are related by a finite sequence of edge flips.*

Now suppose S has a strictly convex real projective structure of finite volume (Ω, Γ, f). An ideal triangulation of S is *straight* if each ideal edge is the image of the intersection of Ω with a projective line. We remark that the word *straight* has been chosen instead of *geodesic* so that the terminology can be transferred to the more general case of properly convex projective structures, where geodesics (with respect to the Hilbert metric) generally do not develop into straight lines in the projective plane.

Lemma 4.3.3. *Suppose S has a strictly convex real projective structure of finite volume. Every ideal triangulation of S is isotopic to a straight ideal triangulation.*

Proof. The ideal triangulation T of S can be lifted to a Γ–equivariant topological ideal triangulation $\widetilde{\mathrm{T}}$ of $\widetilde{S}$ and mapped to Ω using a developing map. There is a homeomorphism of Ω that fixes the frontier $\partial\overline{\Omega}$ and takes each topological ideal edge to a segment of a projective line. Since $\widetilde{S}$ is contractible, no ideal triangle in $\widetilde{S}$ has two edges identified. Given an ideal triangle in $\widetilde{S}$, we claim that the two edges sharing an ideal vertex in $\widetilde{S}$ are mapped to line segments meeting at a point of $\partial\overline{\Omega}$. This follows since the endpoint of an ideal edge must be mapped to a subset of $\partial\overline{\Omega}$ that is invariant under the corresponding peripheral subgroup. Since the structure is of finite volume, this is a single point.

Since $\overline{\Omega}$ is a closed disc, this homeomorphism is isotopic to the identity. Whence the topological ideal triangulation $\widetilde{\mathrm{T}}$ of Ω is isotopic to a straight ideal triangulation. Since the set of ideal endpoints of edges is Γ–equivariant and any two such endpoints determine a unique segment of a projective line in Ω, the straight ideal triangulation is also Γ–equivariant. We may therefore choose a Γ–equivariant isotopy between the ideal triangulations of Ω and push this down to an isotopy of S. ■

Remark 4.3.4. *The above proof used the fact that the structure is of finite volume. In case of a properly convex projective structure, incident ideal edges of an ideal triangles may limit to distinct points on a line segment in $\partial\overline{\Omega}$ connecting two distinct fixed points of the holonomy of a peripheral element. If this happens, after straightening, one can equivariantly isotope the triangulation so that these limiting endpoints of ideal edges both move to the same fixed point.*

In the following, we will use an ideal triangulation T of S to obtain parametrisations of $\mathcal{T}_3^+(S)$ and $\mathcal{T}_3^-(S)$. This requires us to clarify the relationship between the smooth structure on S and the ideal triangulation T. There are two natural viewpoints:

1. In its most general form, a real analytic structure on S is defined by a maximal atlas or an equivalence class of atlases. Thurston [48, p.111] calls this a *polite fiction*. In this case, we appeal to the uniformisation theorem in order to obtain an analytic isomorphism $\widetilde{S} \to \mathbb{D}^2 \subset \mathbb{R}^2 \subset \mathbb{RP}^2$ (see Hubbard [29, 1.8.8]. We may therefore identify $\widetilde{S}$ with $\mathbb{D}^2$, and view this as the hyperbolic plane as given in the Klein model (cf. 2.3.1). Moreover, this gives S a complete hyperbolic metric of finite volume (see [29, 3.3.3]. Given an ideal triangulation T of S we may assume by the above Lemma 4.3.3 that it is straight with respect to the hyperbolic metric. In particular, ideal hyperbolic triangles in $\widetilde{S}$ *are* euclidean triangles in $\mathbb{R}^2$ with their vertices removed and glued via euclidean isometries. We note that two triangles in the same $\pi_1(S)$–orbit in $\widetilde{S}$ are isometric as hyperbolic triangles, but generally not isometric as euclidean triangles.

2. A surface S is given by an ideal triangulation. In this case, we identify each ideal triangle with the regular euclidean triangle of side length one and with its vertices removed, and we realise the face pairings of triangles by euclidean isometries. The resulting identification space has a natural analytic structure, and we assume that S is given with this structure. This construction lifts to the universal cover of S.

To summarise, given a real analytic surface S with an ideal triangulation T, we may assume that each ideal triangle in $\widetilde{S}$ is identified with a euclidean triangle (and hence parameterised using barycentric coordinates), and the face pairings between triangles are realised by euclidean isometries.

Now suppose that $\text{dev}\colon \widetilde{S} \to \Omega \subset \mathbb{RP}^2$ is a developing map for a properly convex projective structure on S. By definition, this is an analytic map, but the images of the ideal edges may not be straight in Ω. Using the methods from the proof of Lemma 4.3.3 and Remark 4.3.4, we may compose dev with an isotopy to obtain a map that is linear on each ideal triangle and continuous. We will term the resulting map a *PL developing map.*

Conversely, suppose Ω is a properly convex domain and $\text{hol}\colon \pi_1(S) \to \text{SL}(\Omega)$ is a discrete and faithful representation. Assume that $\text{dev}'\colon \widetilde{S} \to \Omega \subset \mathbb{RP}^2$ is a continuous bijection that is linear on each ideal triangle and equivariant with respect to hol. Choose a regular neighbourhood N of the union of the ideal edges. We may equivariantly approximate dev' by an analytic map, only changing it in N. This gives a developing map $\text{dev}\colon \widetilde{S} \to \Omega \subset \mathbb{RP}^2$ that agrees with dev' in the complement of N, and hence is uniquely determined by dev'. Moreover, the images of the ideal endpoints of the edges will remain unchanged, and hence applying the straightening procedure to dev again results in dev'.

The upshot of this discussion is that we can work interchangeably with developing maps and PL developing maps, and the latter are uniquely determined by the limiting endpoints of the ideal edges on the boundary of the properly convex domain.

4.4 The canonical bijection for positively framed structures

For a fixed ideal triangulation T of S, let $\triangle$ be the set of all ideal triangles and $\underline{E}$ be the set of all oriented edges. Given a point in $\mathcal{T}_3^+(S)$, to each ideal vertex of the lift of T to the universal cover of S there is both an associated point in the projective plane and a line though that point, a flag. Whence to each ideal triangle there is an associated triple of flags, and hence (after choosing a cyclic order) a triple ratio. Similarly, to each oriented edge, there is an associated quadruple ratio of flags. This gives rise to a map $\mathcal{T}_3^+(S) \to \mathbb{R}_{>0}^{\triangle \cup \underline{E}}$. The proof of the below Theorem will make the association precise and turn it into a well-defined bijective map.

The *Fock-Goncharov moduli space* is then the set of all positive real valued functions $\mathbb{R}_{>0}^{\triangle \cup \underline{E}} := \{\triangle \cup \underline{E} \to \mathbb{R}_{>0}\}$. We note that $|\triangle \cup \underline{E}| = -8\chi(S)$.

Theorem 4.4.1 (Fock-Goncharov 2007, [15]). *Let $S = S_{g,n}$ be a surface of negative Euler characteristic with at least one puncture. For each ideal triangulation* T *of S and each orientation ν on S there is a canonical homeomorphism*

$$\phi^+_{\mathrm{T},\nu} : \mathcal{T}_3^+(S) \to \mathbb{R}_{>0}^{\triangle \cup \underline{E}}.$$

Remark 4.4.2. *It follows from §5.1.3 that choosing different triangulations* T *and* T′ *results in a change of coordinate map $\mathbb{R}_{>0}^{\triangle\cup\underline{E}} \to \mathbb{R}_{>0}^{\triangle\cup\underline{E}}$ that is real analytic. In particular, the bijection $\phi^+_{\mathrm{T},\nu}$ from the above theorem allows us to define a real analytic structure on $\mathcal{T}_3^+(S)$ that is independent of the triangulation.*

The same reasoning applies to $\mathcal{T}_3^-(S)$, and it follows from §5.1.2 that the natural duality maps between $\mathcal{T}_3^+(S)$ and $\mathcal{T}_3^-(S)$ are again real analytic.

By fixing an orientation of S, we may suppress ν from the notation and simply write $\phi^+_{\mathrm{T},\nu} = \phi^+_{\mathrm{T}}$. In the introduction we suppressed both the triangulation and the orientation from the notation and use the inverse $\psi^+ = (\phi^+_{\mathrm{T},\nu})^{-1}$.

We will call $\phi^+_{\mathrm{T}}(x)$ the *Fock-Goncharov coordinate* of $x \in \mathcal{T}_3^+(S)$. Theorem 1.0.4 follows from Theorem 4.4.1 and Remark 4.4.2.

Proof. Assume S is oriented using ν. We first define $\phi^+_{\mathrm{T},\nu} = \phi^+_{\mathrm{T}}$.

Let $(\widetilde{S}, \pi)$ be the universal cover of S and let $\widetilde{\mathrm{T}}$ be the lift of T to $(\widetilde{S}, \pi)$. Let $x = (\mathrm{dev}_x, \mathrm{hol}_x)^+ \in \mathcal{T}_3^+(S)$. The idea is to choose a PL developing map in its isotopy class that is adapted to the framing and has straight ideal triangles, in order to define triple and quadruple ratios.

Fix an affine patch $\mathcal{A}$ in which Ω is strictly contained. We observe that every ideal vertex of $\widetilde{\mathrm{T}}$ is stabilised by a subgroup of $\pi_1(S)$ conjugate to a peripheral subgroup. Hence we choose a developing map in the isotopy class dev_x by mapping each ideal vertex of $\widetilde{\mathrm{T}}$ to a point which is the first entry of some flag of the framing. This choice also determines the images of the edges in $\widetilde{\mathrm{T}}$ as straight line segments between the vertices.

Recall that each point in the framing is paired up with a supporting line. Hence to each ideal triangle we have an associated triple of flags. More specifically, let $t_i \in \triangle$ be an ideal triangle. Any lift $\widetilde{t_i}$ of t_i to $\widetilde{S}$ is assigned a triple of flags $\mathfrak{F}_i := (\!(\mathcal{F}_{i,0}, \mathcal{F}_{i,1}, \mathcal{F}_{i,2})\!)$, cyclically ordered according to the orientation induced on t_i by ν. Since Ω is properly convex, $\mathfrak{F}_i$ represents an element of $\mathcal{P}_3$. Hence define

$$\phi^+_{\mathrm{T}}(x)(t_i) := \mathfrak{Z}(\mathfrak{F}_i) \in \mathbb{R}_{>0}.$$

There is a choice of $\widetilde{t_i}$ in this definition. Any two such choices differ by a deck transformation. Since dev_x is hol_x–equivariant, any two resulting triples of flags $\mathfrak{F}_i$ and $\mathfrak{F}_i'$ differ by a projective transformation and therefore represent the same element of $\mathcal{P}_3$. Lemma 3.2.1 ensures that $\mathfrak{Z}(\mathfrak{F}_i) = \mathfrak{Z}(\mathfrak{F}_i')$.

Let $e_i \in \underline{E}$ be an oriented edge. Choose a lift $\widetilde{e_i}$ of e_i. $\widetilde{e_i}$ is shared by two triangles in $\widetilde{\mathrm{T}}$, each of which is assigned a triple of flags. Thereby we can consider $\widetilde{e_i}$ to be uniquely associated to an ordered quadruple of flags $\mathfrak{F}_i := (\mathcal{F}_{i,0}, \mathcal{F}_{i,1}, \mathcal{F}_{i,2}, \mathcal{F}_{i,3})$, ordered according to the orientation of Ω, where $\mathcal{F}_{i,0}$ and $\mathcal{F}_{i,2}$ represent the vertices at the tail and head of $\mathrm{dev}_x(\tilde{e_i})$ respectively. As Ω is properly convex, $\mathfrak{F}_i$ is identified with an element of $\mathcal{P}_4^*$. Thereby we define

$$\phi_{\mathrm{T}}^+(x)(e_i) := \not{A}(\mathfrak{F}_i) \in \mathbb{R}_{>0}.$$

As above, another choice of $\tilde{e_i}$ only changes $\mathfrak{F}_i$ by a projective transformation and does not change $\not{A}(\mathfrak{F}_i)$. The positivity of $\phi_{\mathrm{T}}^+(x)(e_i)$ follows from Theorem 3.7.2.

The above construction gives rise to a well-defined map

$$\phi_{\mathrm{T}}^+ : \mathcal{T}_3^+(S) \to \mathbb{R}_{>0}^{\triangle \cup E}.$$

It remains to show that this map is a bijection.

Let $x, y \in \mathcal{T}_3^+(S)$ and $\phi_{\mathrm{T}}^+(x) = \phi_{\mathrm{T}}^+(y)$. For any triangle $t \in \triangle$,

$$\phi_{\mathrm{T}}^+(x)(t) = \phi_{\mathrm{T}}^+(y)(t).$$

Fix an arbitrary lift $\tilde{t}$ of t to $\tilde{S}$. Then $\tilde{t}$ is assigned triples of flags $\mathfrak{F}_x^{\tilde{t}}$ and $\mathfrak{F}_y^{\tilde{t}}$ by the framings of x and y respectively. By Theorem 3.7.1, there exists $A \in \mathrm{SL}(3, \mathbb{R})$ such that

$$A \cdot \mathfrak{F}_x^{\tilde{t}} = \mathfrak{F}_y^{\tilde{t}}.$$

Following Lemma 4.3.3, we assume that edges of ideal triangles are segments of projective lines, which are thus uniquely determined by $\mathfrak{F}_x^{\tilde{t}}$ and $\mathfrak{F}_y^{\tilde{t}}$. Therefore $A \cdot \mathrm{dev}_x(\tilde{t}) = \mathrm{dev}_y(\tilde{t})$. In particular, the restrictions of $A \cdot \mathrm{dev}_x$ and dev_y to $\tilde{t}$ are the same up to an isotopy of their image that fixes the vertices.

We claim that $A \cdot \Omega_x = \Omega_y$ and furthermore that $A \cdot \mathrm{dev}_x$ is isotopic to dev_y over the whole domain. Let t' be a triangle adjacent to t and $\tilde{t}'$ the lift of t' which is adjacent to $\tilde{t}$. Let $\tilde{t}$ and $\tilde{t}'$ share the edge e and let v' be the ideal vertex of $\tilde{t}'$ not contained in e. By proper convexity of Ω_x and Ω_y, the flags at the vertices of $\tilde{t}$ and $\tilde{t}'$ form two pairs of strictly inscribed convex quadrilaterals χ_x and χ_y. Cyclically order them according to ν, and endow them with the same marking by choosing an endpoint of e. Thus they are elements of $\mathcal{P}_4^*$ having same triple ratios and quadruple ratios. By Theorem 3.7.2, χ_x and χ_y are projectively equivalent. Since $A \cdot \mathrm{dev}_x(\tilde{t}) = \mathrm{dev}_y(\tilde{t})$, it is also the case that $A \cdot \chi_x = \chi_y$. In particular the projective lines η_x' and η_y' at $\mathrm{dev}_x(v')$ and $\mathrm{dev}_y(v')$ respectively, are projectively equivalent via A. Lemma 4.3.3 ensures once again that $A \cdot \mathrm{dev}_x(\tilde{t}') = \mathrm{dev}_y(\tilde{t}')$, and $A \cdot \mathrm{dev}_x$ and dev_y are isotopic on $\tilde{t} \cup \tilde{t}'$.

Continuing in this manner we see that the image of $\tilde{\mathrm{T}}$ is uniquely determined by $\phi_{\mathrm{T}}^+(x) = \phi_{\mathrm{T}}^+(y)$. Moreover, since both developing maps are equivariant with respect to the action of their corresponding holonomy groups, we have

$$\mathrm{hol}_x(\gamma) = A^{-1} \cdot \mathrm{hol}_y(\gamma) \cdot A, \quad \forall\, \gamma \in \pi_1(S).$$

Therefore $A \cdot \mathrm{dev}_x$ and dev_y differ by a holonomy-equivariant isotopy, which may be pushed down to S. Since the framing is determined by the flags at vertices of the ideal triangles, we can conclude that x and y define equivalent framed convex projective structures on S. This completes the proof that ϕ^+_{T} is injective.

It remains to show that ϕ^+_{T} is surjective. Suppose each triangle and each oriented edge in T were assigned a positive real number. Lift these assignments to $\widetilde{\mathrm{T}}$ equivariantly. We will explicitly construct a domain $\Omega^+ \subset \mathbb{RP}^2$ and holonomy representation $\mathrm{hol} : \pi_1(S) \to \mathrm{SL}(3, \mathbb{R})$ such that $\Omega^+/\,\mathrm{hol}(\pi_1(S))$ is a convex projective surface homeomorphic to S. Finally we will show that the original assignment of numbers to T determines a canonical choice of framing, so that the convex projective structure on S thus defined is extended to an element of $\mathcal{T}_3^+(S)$.

Let $t \in \Delta$. Fix a lift $\tilde{t}$ of t to $\widetilde{S}$ with orientation inherited from S. By Theorem 3.7.1, the positive real number assigned to $\tilde{t}$ uniquely determines an element of $\mathcal{P}_3$. Hence fix a cyclically ordered representative $\mathfrak{F}^{\tilde{t}}$. This triple of flags defines three points of $\partial\overline{\Omega^+}$ and a triple of supporting lines to Ω^+ at those points. The rest of the construction of Ω^+ proceeds as in the proof of Theorem 3.7.2 where, given a triple of flags, a fourth flag of an adjacent triangle is uniquely determined by two edge ratios and a triple ratio. Iterating this procedure, Ω^+ is defined to be the union of the inscribed triangles determined by all these elements of $\mathcal{P}_3$, and the flags will be the framing for the projective structure.

We describe Ω^+ in more detail. Since the set of triangles of $\widetilde{\mathrm{T}}$ is countable, we can index them as $\{\tilde{t}_i\}_{i \in \mathbb{N}}$ so that $\cup_{i=0}^m t_i$ is connected for all $m \in \mathbb{N}$, and $\cup_{i=0}^\infty \tilde{t}_i = \widetilde{\mathrm{T}}$. In the above construction we assigned to each triangle $\tilde{t}_i \in \widetilde{\mathrm{T}}$ an element χ_i of $\mathcal{P}_3$. Let $\mathcal{T}_i$ be the inscribed triangle of χ_i with the vertices removed, thus define

$$\Omega^+_m := \bigcup_{i=0}^{m} \mathcal{T}_i \qquad \text{and} \qquad \Omega^+ := \bigcup_{i=0}^{\infty} \mathcal{T}_i.$$

We are going to show that Ω^+ is properly convex. Since Ω^+_0 is clearly properly convex, we will proceed by induction on m. Suppose Ω^+_{m-1} is properly convex and fix an affine patch $\mathcal{A}$ strictly containing Ω^+_{m-1}. Let e be the common edge between Ω^+_{m-1} and $\mathcal{T}_m$ and let V be the vertex of $\mathcal{T}_m$ disjoint from e. The edge e is equipped with two edge ratios and a flag (V_i, η_i) for each endpoint labelled by $i = 0, 1$. We remark that η_i is a supporting line for the domain Ω^+_{m-1}.

Let Δ_m be the triangle defined by η_0, η_1 and e which is disjoint from Ω^+_{m-1}. If necessary, we choose a new patch $\mathcal{A}$ that contains both Δ_m and Ω^+_{m-1}. If $\mathcal{A}$ is not already such a patch then this choice is possible because $\overline{\Omega^+_{m-1}}$ does not contain $\eta_0 \cap \eta_1$. We may therefore choose a new affine patch by choosing the complement of a line which passes 'close' to $\eta_0 \cap \eta_1$ and does not intersect Δ_m. If it were the case that $\overline{\Omega^+_{m-1}}$ contained $\eta_0 \cap \eta_1$, then by proper convexity Ω^+_{m-1} would be a triangle, hence $m = 1$ and the triple of flags used to define χ_0 would not have been in general position.

Since the edge ratios are strictly positive, Theorem 3.7.2 implies that V lies strictly within the triangle Δ_m. It follows that Ω_m^+ is also convex and properly contained in the affine patch $\mathcal{A}$. To conclude the induction argument it is enough to observe that the positivity of the triple ratio associated to $\mathcal{T}_m$ implies that η_V, the line attached to V, is also a supporting line for Ω_m^+. Therefore $\Delta_{m+1} \subset \Delta_m \subset \Delta_2$. In particular Ω_m^+ is strictly contained in the circumscribed triangle of χ_1 for all m. Hence Ω^+ is properly convex.

The set Ω^+ is defined as a union of ideal triangles, thus it has a natural combinatorial structure. By construction, we have a piecewise linear homeomorphism from the universal cover $\widetilde{S}$ to Ω^+, and we now define an action of $\pi_1(S)$ by projective maps on Ω^+ that has this piecewise linear homeomorphism as a PL developing map. That is the holonomy representation hol : $\pi_1(S) \to \Gamma < \mathrm{SL}(3,\mathbb{R})$. Fix an arbitrary lift $\tilde{t}$ of some $t \in \triangle$. Let $\mathfrak{F}^{\tilde{t}}$ be the cyclically ordered triple of flags associated to $\tilde{t}$. For each $\gamma \in \pi_1(S)$, $\tilde{t}' := \gamma(\tilde{t})$ is another lift of t, which is assigned another triple of flags $\mathfrak{F}^{\tilde{t}'}$ with the same triple ratio as $\mathfrak{F}^{\tilde{t}}$. Once again, Theorem 3.7.1 applies to provide a projective transformation $T_\gamma \in \mathrm{SL}(3,\mathbb{R})$ such that

$$T_\gamma \cdot \mathfrak{F}^{\tilde{t}} = \mathfrak{F}^{\tilde{t}'}.$$

Uniqueness of T follows from Lemma 3.1.1. Now T_γ is an element of $\mathrm{SL}(\Omega^+)$, namely it preserves Ω^+. Indeed, Ω^+ is defined exclusively in terms of triangle parameters and edge parameters, which are invariant under projective transformations. More precisely, given two triangles $\tilde{t}, \tilde{t}' \subset \widetilde{\mathrm{T}}$,

$$\tilde{t} = \gamma(\tilde{t}') \iff \mathrm{dev}(\tilde{t}) = T_\gamma \cdot \mathrm{dev}(\tilde{t}').$$

Hence we have

$$\mathrm{hol}(\gamma) := T_\gamma, \quad \forall\, \gamma \in \pi_1(S).$$

The PL developing map is equivariant with respect to hol: $\gamma \mapsto T_\gamma$, which is an isomorphism onto its image $\Gamma := \mathrm{hol}(\pi_1(S))$. In particular, dev induces a PL homeomorphism f from S to Ω^+/Γ. The triple (Ω^+, Γ, f) thereby defines an element of $\mathcal{T}_3(S)$.

The next step is to show that $(\Omega^+, \Gamma, f) \in \mathcal{T}_3^m(S)$ and that it has a natural framing given by the flags at each ideal vertex. Before we analyse maximal and minimal ends, we characterise flags supporting of Ω^+ as fixed elements of conjugates of peripheral subgroups of Γ.

Recall that every ideal vertex V of Ω^+ has a line η assigned to it, forming the flag (V, η). As dev is a combinatorial equivalence, V corresponds to an ideal vertex $\widetilde{\mathcal{E}}$ of $\widetilde{\mathrm{T}}$, which is a lift of an end $\mathcal{E}$ of S. Let γ_0 be a generator of $\pi_1(\mathcal{E})$, and γ the element conjugate to γ_0 fixing $\widetilde{\mathcal{E}}$. Then V is fixed by $\mathrm{hol}(\gamma)$. Furthermore, if $\tilde{t}$ is a triangle in Ω^+ with ideal vertex V, then $\mathrm{hol}(\gamma)(\tilde{t})$ is, by construction, another triangle of Ω^+ sharing the same flag (V, η). Therefore η is also preserved by $\mathrm{hol}(\gamma)$. Summarising, every flag supporting Ω^+ is a pair consisting of a fixed point and an invariant line of some (conjugate of a) peripheral subgroup of Γ. When $\mathcal{E}$ is a special end or a cusp, the flag (V, η) is an admissible choice of positive framing at $\mathcal{E}$.

Now suppose that $\mathcal{E}$ is hyperbolic, namely $\mathrm{hol}(\gamma)$ fixes three distinct points of $\mathbb{RP}^2$, one attracting V_+, one repelling V_-, and a saddle V_0. By the previous discussion there are six possibilities for (V, η), such that

$$V \in \{V_+, V_-, V_0\}, \quad \eta \in \{V_+V_-, V_+V_0, V_-V_0\} \quad \text{and} \quad V \in \eta.$$

For convenience, we fix an affine patch $\mathcal{A}$ strictly containing Ω^+. Recall from §2.2.8 that the interior of the segment between V_+ and V_- in $\mathcal{A}$, say s, is either contained in Ω^+ or in the frontier of Ω^+, and Ω^+ can not be equal to the interior of a triangle bounded by s and V_0.

If $V = V_0$, then both V_0 and s are subsets of $\overline{\Omega^+}$. By proper convexity, Ω^+ must contain the interior of the triangle spanned by V_+, V_-, V_0, and therefore $\mathcal{E}$ must be maximal.

Suppose $V \neq V_0$. Let t be an ideal triangle of Ω^+ with ideal vertex V. If s was contained in Ω^+, then by the construction s would be covered by some triangles of Ω^+. However, the orbit of t under $\langle \mathrm{hol}(\gamma) \rangle$ accumulates at s, so s can not intersect the interior of any triangle. At the same time, s can not be the edge of a triangle, as $\mathrm{hol}(\gamma)$ acts non-trivially on s. It follows that s must be in the frontier of Ω^+, and $\mathcal{E}$ is minimal.

In conclusion, every hyperbolic end $\mathcal{E}$ is either maximal or minimal, and (V, η) is an admissible choice of positive framing also at $\mathcal{E}$. Framing each end in this way gives $(\Omega^+, \Gamma, f)^+ \in \mathcal{T}_3^+(S)$.

To conclude the proof, we need to show that the bijection that we have described is a homeomorphism between $\mathcal{T}_3^+(S)$ with the weak topology on developing maps and $\mathbb{R}_{>0}^{\Delta \cup E}$ with the usual topology. A compact set in $\widetilde{S}$ only meets finitely many ideal triangles. A small perturbation of a PL developing map results in a small perturbation of the images of peripheral fixed points and supporting lines. In particular, one obtains a small perturbation of the associated Fock-Goncharov coordinates. To show that the inverse map is continuous, we need control over the PL developing maps constructed for a given Fock-Goncharov coordinate. This arises by fixing the developing image of a given decorated base triangle in $\widetilde{S}$—a particularly nice choice, coined cloverleaf position, is described in §4.8. ■

4.5 A natural action of the symmetric group

The *symmetric group* on three letters, $\mathrm{Sym}(3)$, acts by permuting the set of framings of a given end. This induces a $\mathrm{Sym}(3)^n$–action on $\mathcal{T}_3^+(S)$.

Theorem 4.5.1. *Let $S = S_{g,n}$ be a surface of negative Euler characteristic with $n > 0$ punctures. There is a faithful but not free action $\mathrm{Sym}(3)^n \curvearrowright \mathcal{T}_3^+(S)$, whose orbits are framed convex projective structures having conjugate holonomy representations. In particular, the map $\mu^+ \colon \mathcal{T}_3^+(S) \to \mathfrak{X}^\times(S)$ is an isomorphism and the quotient space $\mathcal{T}_3^+(S)/\mathrm{Sym}(3)^n$ is naturally identified with $\mathfrak{X}(S)$.*

Proof. First we construct a Sym(3)–action on the moduli space of a once punctured surface $S = S_{g,1}$. Denote the end of S by $\mathcal{E}$ and let $\gamma \in \pi_1(S)$ be a generator of the peripheral subgroup. Every point $x = (\Omega^x, \Gamma^x, f^x)^+ \in \mathcal{T}_3^+(S)$ comes with a holonomy representation hol_x and a Γ^x–orbit of flags constituting the framing of $\mathcal{E}$. Let (V, η) be the unique element in that orbit such that V is fixed by $\mathrm{hol}_x(\gamma)$. As a projective transformation, $\mathrm{hol}_x(\gamma)$ fixes three points (counted with multiplicity) V_1, V_2 and V_3 in $\mathbb{RP}^2$. We order them so that:

1. $V_1 = V$,
2. $V_2 \in \eta$,
3. if $\mathcal{E}$ is special, $V_1 = V_2$ if and only if $V_3 \notin \eta$.

Thereby, x is assigned the Γ^x–orbit of an ordered triple (V_1, V_2, V_3). For $\alpha \in \mathrm{Sym}(3)$, we define $y := \alpha \cdot x$ to be that structure whose assigned Γ^x–orbit of ordered triple is $(V_{\alpha(1)}, V_{\alpha(2)}, V_{\alpha(3)})$. If $\mathcal{E}$ is not hyperbolic, the structure y has equivalent Hilbert geometry to $\pi^+(x) = (\Omega^x, \Gamma^x, f^x)$, but it is framed according to the conditions 1) – 3) above, applied to the triple $(V_{\alpha(1)}, V_{\alpha(2)}, V_{\alpha(3)})$. If $\mathcal{E}$ is hyperbolic, one may need to retract or expand Ω^x to match the correct framing. In this case, the flag $(V_{\alpha(1)}, V_{\alpha(1)}V_{\alpha(2)})$ uniquely determines whether $\mathcal{E}$ is maximal or minimal, hence define Ω^y accordingly. Then Ω^x/Γ^x differs from Ω^y/Γ^x by at most an annulus, so there is a natural homeomorphism $g : \Omega^x/\Gamma^x \to \Omega^y/\Gamma^x$ which is the identity outside a sufficiently small regular neighbourhood of $\mathcal{E}$. We define $f^y := f^x \circ g$ and y is the structure $(\Omega^y, \Gamma^x, f^y)$, framed as per $(V_{\alpha(1)}, V_{\alpha(2)}, V_{\alpha(3)})$. We remark that x and y have the same holonomy representations.

The above construction extends to the case $n \geqslant 1$, by acting on a different boundary component with each distinct copy of Sym(3). The stabilizer of a structure with all ends hyperbolic is trivial, therefore the action is faithful. Structures with only cuspidal ends are global fixed points, hence the action is not free. The existence of both types of structures follows by a standard construction in hyperbolic geometry.

It remains to show that the $\mathrm{Sym}(3)^n$–orbit of a point x is precisely the set of those structures in $\mathcal{T}_3^+(S)$ with holonomy conjugate to x. One inclusion is trivial as the Sym(3)–action preserves the holonomy. For the other, suppose $x = (\Omega^x, \Gamma^x, f^x)^+$ and $y = (\Omega^y, \Gamma^y, f^y)^+$ have conjugate holonomies. By acting with the appropriate element of $\mathrm{SL}(3, \mathbb{R})$, we can fix a representative of y which has the same holonomy as x. It follows that $\Gamma^x = \Gamma^y$ and they share the same peripheral subgroups, hence have the same fixed points in $\mathbb{RP}^2$. In particular, for each end $\mathcal{E}_k$ of S there is an element $\alpha_k \in \mathrm{Sym}(3)$ mapping the ordered triple associated to the framing of x at $\mathcal{E}_k$ to the ordered triple associated to the framing of y at $\mathcal{E}_k$. This is enough to prove that Ω^x and Ω^y may only differ at the hyperbolic ends, which can be maximal or minimal.

In particular, analogous to the construction of the developing map in the proof of Theorem 4.4.1, the straightening procedure of ideal triangulation can now be applied to convex cores of the structures x and y respectively, to give a homeomorphism

$g : \Omega^x/\Gamma^x \to \Omega^y/\Gamma^y$ which is the identity outside a small neighbourhood of $\cup_k \mathcal{E}_k$, such that $f^y := f^x \circ g$. This concludes the proof that y is indeed $\alpha \cdot x$, for some $\alpha = (\alpha_1, \ldots, \alpha_n) \in \mathrm{Sym}(3)^n$.

Associating flags to the holonomies gives the claim regarding the map to $\mathfrak{X}^\times(S)$. ■

Global fixed points of the action are those structures whose ends are all cusps. In §6.1 we show that this set is in fact $\mathcal{T}_3^f(S)$. In §5.3 we give a parametrisation of a fundamental domain for the action of $\mathrm{Sym}(3)^n$.

4.6 Duality

Every convex projective structure has a dual structure, constructed using the self-duality of the projective plane. This induces an isomorphism $\sigma^+ \colon \mathcal{T}_3^+(S) \to \mathcal{T}_3^-(S)$. In this section we recall the main points of this construction, and refer the reader to [20, 50, 9] for details and proofs.

Let Ω be a properly convex domain of $\mathbb{RP}^2$. Its *dual domain* is the set

$$\Omega^* := \{P \in \mathbb{RP}^2 \mid P^\perp \cap \overline{\Omega} = \varnothing\}.$$

Then Ω^* is also a properly convex domain and $\Omega = (\Omega^*)^*$. Supporting lines to Ω correspond to points on the frontier of Ω^*, and vice-versa. Furthermore, there is a real analytic diffeomorphism $\Phi : \Omega \to \Omega^*$, called the *dual map*. The map Φ has an elementary interpretation as follows. For $P \in \Omega$, the line $P^\perp$ is disjoint from $\overline{\Omega^*}$, thus the affine patch $\mathcal{A}_P = \mathbb{RP}^2 \backslash P^\perp$ strictly contains Ω^*. Then $\Phi(P)$ is the centre of mass of Ω^* in $\mathcal{A}_P$.

In general, Φ is not a projective transformation, unless Ω is a conic. However, for $A \in \mathrm{PGL}(\Omega)$ and $P \in \Omega$, $\Phi(A(P)) = (A^t)^{-1}(\Phi(P))$. Hence we define the *dual group* of a subgroup $\Gamma \leqslant \mathrm{PGL}(3, \mathbb{R})$ to be the group

$$\Gamma^* := \{(A^t)^{-1} \mid A \in \Gamma\}.$$

Clearly $\Gamma \cong \Gamma^*$. The upshot of this summary of duality is the following result.

Theorem 4.6.1. *Let $(\Omega, \Gamma, f) \in \mathcal{T}_3(S)$. Then Φ induces a real analytic diffeomorphism*

$$\overline{\Phi} : \Omega/\Gamma \to \Omega^*/\Gamma^*.$$

In particular, $(\Omega^*, \Gamma^*, f^*)$ is an element of $\mathcal{T}_3(S)$, where $f^* := \overline{\Phi} \circ f$.

Theorem 4.6.1 defines an involution $\sigma \colon \mathcal{T}_3(S) \to \mathcal{T}_3(S)$, taking a structure to its dual structure. Duality inverts the eigenvalues of the holonomy, therefore $\overline{\Phi}$ preserves cusps and maps hyperbolic (resp. quasi-hyperbolic) ends to hyperbolic (resp. quasi-hyperbolic) ends. However it reverses inclusions,

$$\Omega_1 \subset \Omega_2 \iff \Omega_1^* \supset \Omega_2^*.$$

Hence a hyperbolic end $\mathcal{E}$ is maximal (resp. minimal) for a structure x if and only if it is minimal (resp. maximal) for $\sigma(x)$. It follows that σ restricts to $\mathcal{T}_3^m(S)$. When $x \in \mathcal{T}_3^m(S)$ is enriched with a positive framing, its dual structure $\sigma(x)$ has a natural negative framing. Specifically, suppose (V, η) is a flag in the Γ–orbit of the positive framing of an end $\mathcal{E}$, fixed by an element $\mathrm{hol}(\gamma)$. Then $(\eta^\perp, V^\perp)$ is fixed by $(\mathrm{hol}(\gamma)^t)^{-1}$ and its Γ^*–orbit is an admissible dual negative framing of the same end $\mathcal{E}$. In particular, if V_0, V_+, V_- and W_0, W_+, W_- are the saddle, attracting and repelling fixed points of $\mathrm{hol}(\gamma)$ and $(\mathrm{hol}(\gamma)^t)^{-1}$ respectively, the dual relations between them is

$$\begin{aligned} V_+^\perp &= W_-W_0, & V_-^\perp &= W_+W_0, & V_0^\perp &= W_+W_-, \\ (V_+V_-)^\perp &= W_0, & (V_-V_0)^\perp &= W_+, & (V_+V_0)^\perp &= W_-. \end{aligned}$$

Therefore σ lifts to an isomorphism $\sigma^+ \colon \mathcal{T}_3^+(S) \to \mathcal{T}_3^-(S)$, that we call the *duality map*. The above argument gives a natural isomorphism $\sigma^- \colon \mathcal{T}_3^-(S) \to \mathcal{T}_3^+(S)$ which is inverse to σ^+.

It follows from this discussion, that the composition $\phi^+_{\mathrm{T},\nu} \circ \sigma^- \colon \mathcal{T}_3^-(S) \to \mathbb{R}_{>0}^{\Delta \cup E}$ gives a natural bijection. However, the associated conjugacy classes of framed holonomies for the same point in $\mathbb{R}_{>0}^{\Delta \cup E}$ will be projectively dual to each other. This can be remedied by *defining* the map $\phi^-_{\mathrm{T},\nu}$ as

$$\phi^+_{\mathrm{T},\nu} \circ (\mu^+)^{-1} \circ \sigma \circ \mu^+ \circ \sigma^{-1},$$

where $\sigma \colon \mathfrak{X}^\times(S) \to \mathfrak{X}^\times(S)$ takes the character of the representation ρ to the character of the inverse transpose of the representation, and each flag to the dual flag. We *construct* the map $\phi^-_{\mathrm{T},\nu}$ in this way in the next section.

4.7 The canonical bijection for negatively framed structures

In §2.2.8 we defined the negative framing of a convex projective structure and the moduli space $\mathcal{T}_3^-(S)$. The proof of the following result shows how to parametrise $\mathcal{T}_3^-(S)$ in a manner analogous to that of $\mathcal{T}_3^+(S)$. Moreover, as in Remark 4.4.2, we may combine Theorem 4.7.1 with the description of the change of coordinate functions in §5.1.3 to imbue $\mathcal{T}_3^-(S)$ with a real analytic structure.

Theorem 4.7.1. *Let $S = S_{g,n}$ be a surface of negative Euler characteristic with at least one puncture. For each ideal triangulation* T *of S and each orientation ν on S there is a canonical isomorphism*

$$\phi^-_{\mathrm{T},\nu} : \mathcal{T}_3^-(S) \to \mathbb{R}_{>0}^{\Delta \cup E}.$$

Once again, we fix an orientation of S, and suppress ν from the notation to simply write $\phi^-_{\mathrm{T},\nu} = \phi^-_{\mathrm{T}}$. The map in the introduction is $\psi^- = (\phi^-_{\mathrm{T},\nu})^{-1}$.

Proof. One can follow the proof of Theorem 4.4.1 verbatim until the end of the proof of injectivity. We recall that ϕ_{T}^{+} was defined by mapping the lift to $\widetilde{S}$ of the ideal triangulation of S into Ω with vertices at the peripheral fixed points in the frontier of Ω, and then taking triple ratios and quadruple ratios of the appropriate flags associated to edges and vertices in the triangulation. In the proof of the injectivity, it turned out that $\Omega = \Omega^+$, the domain obtained as the image of the ideal triangulation of $\widetilde{S}$.

The same construction is now applied to define ϕ_{T}^{-}, which produces a point in $\mathbb{R}_{>0}^{\triangle\cup E}$ associated to an element of $\mathcal{T}_3^-(S)$. However, in certain cases, the negatively framed domain Ω strictly contains the image Ω^+ of $\widetilde{S}$. This happens precisely when we frame a maximal hyperbolic end with one of the two non-saddle points, and the line through the saddle point. Then $\Omega\backslash\Omega^+$ consists of the orbits of maximal cusps framed in this way. One can now produce an element of $\mathcal{T}_3^-(S)$ by thickening the structure at these ends. Alternatively, one may appeal to duality as follows, which in particular proves the claim that the new domain is the intersection of the circumscribing triangles.

Referring to the setting and the notation in the proof of Theorem 4.4.1, we define $\mathcal{T}_i^c$ to be the circumscribed triangle in χ_i (containing the inscribed triangle). Then

$$\Omega^- := \mathrm{Int}\Big(\bigcap_{i=0}^{\infty}\mathcal{T}_i^c\Big).$$

In this case, Ω^- is clearly properly convex as it is the countable intersection of properly convex sets. We have the action of Γ on Ω^-, and are required to describe a homeomorphism $f\colon S \to \Omega^-/\Gamma$. In the proof of Theorem 4.4.1 this followed from the combinatorial map between the universal cover of S with the induced triangulation to the triangulated domain. However, here Ω^- does not have such a triangulation.

By duality, $(\Omega^-)^*$ is a properly convex domain, union of the triangles $\mathcal{T}_i^*$. Following the discussion in the proof of Theorem 4.4.1, the dual domain $(\Omega^-)^*$ is combinatorially isomorphic to $\widetilde{\mathrm{T}}$, and there is the dual group Γ^* of $\mathrm{PGL}((\Omega^-)^*)$ acting as a simplicial isomorphism on $(\Omega^-)^*$. In particular, $(\Omega^-)^*/\Gamma^* \cong S$ and Theorem 4.6.1 therefore gives a homeomorphism $f\colon S \to (\Omega^-)^*/\Gamma^* \to \Omega/\Gamma$. Whence (Ω, Γ, f) with the given framing is an element of $\mathcal{T}_3^-(S)$. ■

The same Fock-Goncharov coordinate (for a fixed triangulation and orientation of S) gives structures in both $\mathcal{T}_3^+(S)$ and $\mathcal{T}_3^-(S)$. The respective domains Ω^+ and Ω^- for these structures only differ (up to projective equivalence) at those hyperbolic ends whose flags comprise one of the two non-saddle points and the line through the saddle point. In fact, in these cases, Ω^- contains the triangle spanned by the three peripheral fixed points, hence the end is maximal, while Ω^+ does not.[1] However, the proofs of Theorems 4.4.1 and 4.7.1 show that the structures have equivalent framed holonomies, as the developing maps and framed domains are constructed from equivalent sets of inscribed and circumscribed triangles. We record this in the

[1]We note that this distinction does not arise in [15]. Indeed, the proof of Theorem 2.5 in [15] claims that Ω^+ and Ω^- are always equal domains.

next result.

Corollary 4.7.2. *The following diagram commutes:*

$$\begin{array}{ccccc} & & \mathcal{T}_3^+(S) & & \\ & \overset{(\phi^+_{\mathrm{T},\nu})^{-1}}{\nearrow} & & \overset{\mu^+}{\searrow} & \\ \mathbb{R}^{\triangle\cup E}_{>0} & & & & \mathfrak{X}^\times(S). \\ & \underset{(\phi^-_{\mathrm{T},\nu})^{-1}}{\searrow} & & \underset{\mu^-}{\nearrow} & \\ & & \mathcal{T}_3^-(S) & & \end{array}$$

4.8 Cloverleaf position

A Benzécri patch allows the study of continuous families of marked properly convex projective structures. However, given a continuous family of structures, one needs to determine a continuous family of projective transformations that achieve Benzécri position. The investigations in [25] using Fock-Goncharov coordinates led to another normalisation with developing maps varying over a fixed compact domain. This is called *cloverleaf position.* Figure 1.1 shows the images of some developing maps for convex projective structures on the once-punctured torus in cloverleaf position.

Let $S = S_{g,n}$ be a surface of negative Euler characteristic with at least one puncture. Let T be an ideal triangulation of S. Let $\tilde{t}$ and $\tilde{e}$ be an adjacent triangle and edge in the lift $\widetilde{\mathrm{T}}$ of T to the universal cover $\widetilde{S}$.

Given an element in $\mathcal{T}_3^+(S)$, we may choose a representative developing map $\mathrm{dev}\colon \widetilde{S} \to \mathbb{RP}^2$ for this element such that:

1. The ideal vertices of $\tilde{e}$ are mapped to the basis vectors $[1 : 0 : 0]^\intercal$ and $[0 : 1 : 0]^\intercal$;
2. The remaining ideal vertex of $\tilde{t}$ is mapped to the basis vector $[0 : 0 : 1]^\intercal$;
3. The lines $[0 : t_{012} : 1]$, $[1 : 0 : t_{012}]$ and $[t_{012} : 1 : 0]$ are tangent to the frontier of $\Omega = \mathrm{dev}(\widetilde{S})$ (at the images of the vertices of $\tilde{t}$), where t_{012} is the parameter associated to $\tilde{t}$.

Finally, let $\mathcal{A}$ be the affine patch $x + y + z = 1$.

Then $\mathcal{A}$ is the *cloverleaf chart* of the structure on S with respect to $\tilde{t}$ and $\tilde{e}$. Choose standard coordinates $\mathcal{A} = \mathbb{R}^2$. The proof of the next results shows how to achieve the *cloverleaf position* with respect to $\tilde{t}$ and $\tilde{e}$ within this patch.

Theorem 4.8.1 (Haraway-Tillmann 2018, [25]). *Let Ω' be the image of Ω under the affine isomorphism α between a cloverleaf chart and $\mathbb{R}^2 = \mathbb{C}$ that sends the vertices of $\tilde{t}$ to the cube roots of unity and the vertices of $\tilde{e}$ to the primitive cube roots of unity.*

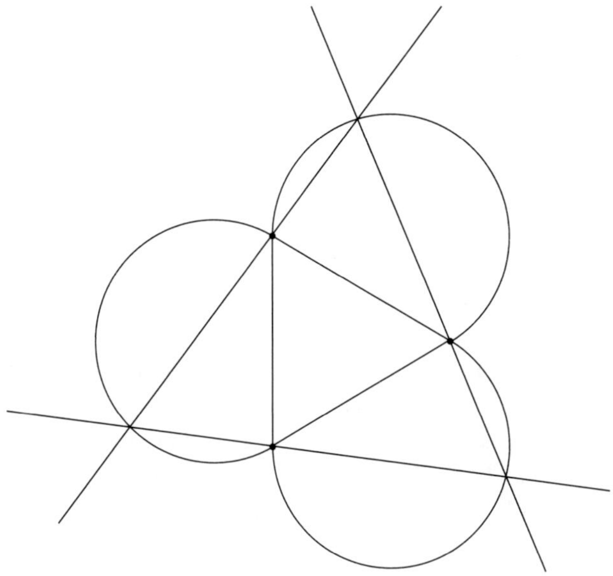

Figure 4.4: Three vertices and supporting hyperplanes in a cloverleaf chart.

Then Ω' contains the triangle whose vertices are the cube roots of unity, and is contained in the union of unit discs centered at the cube roots of -1.

Proof. By definition, the domain Ω' contains the image of $\tilde{t}$, and the image of $\tilde{t}$ under α is the triangle whose vertices are the roots of unity. This concludes the proof of the first claim.

Let

$$\sigma = \begin{pmatrix} 0 & 1 & 0 \\ 0 & 0 & 1 \\ 1 & 0 & 0 \end{pmatrix}.$$

Then σ permutes the image of the vertices of $\tilde{t}$. But it also permutes the lines $\eta_0 = [0 : t_{012} : 1]$, $\eta_1 = [1 : 0 : t_{012}]$ and $\eta_2 = [t_{012} : 1 : 0]$ that are tangent to $\partial\overline{\Omega}$ at the images of the vertices of $\tilde{t}$. So Ω lies within the triangle Δ bounded by η_i. This triangle is symmetric under σ, so its image is symmetric under $s = \alpha \circ \sigma \circ \alpha^{-1}$. But s is an order 3 affine automorphism of $\mathbb{R}^2$ permuting the images of the vertices of $\tilde{t}$, the cube roots of unity. So s is just a $2\pi/3$–rotation about the origin. Therefore the image of Δ is an equilateral triangle circumscribed about the equilateral triangle formed by the roots of unity.

Using elementary Euclidean geometry, it is easy to see that this triangle must lie in the region defined by the union of the unit discs centered at cube roots of -1 (see Figure 4.4). Since Ω' lies inside this triangle, it lies inside the region as well. ■

Chapter 5

Properties of the Parametrisation

5.1 Change of coordinates

Let S be a surface as in Theorem 4.4.1. Having fixed a triangulation T and orientation ν, the space $\mathcal{T}_3^+(S)$ can be canonically parametrised by positive real numbers. A different choice of ν or T may be interpreted as a change of coordinates. Similarly, the duality map $\sigma : \mathcal{T}_3^+(S) \to \mathcal{T}_3^-(S)$, defined by taking a structure to its dual, induces an involution on Fock-Goncharov moduli space. We explicitly construct these transition maps.

We remark that also the action of $\mathrm{Sym}(3)^n$ on $\mathcal{T}_3^+(S)$ descends to a change of coordinates. However, even in the simplest cases, the transition functions are quite convoluted and we do not have a local description of them.

5.1.1 Transition maps for a different orientation

The transition map associated to a switch in the orientation of S is simple to describe. Denote by $-\nu$ the opposite orientation of ν. Then for all $q \in \triangle \cup \underline{E}$,

$$\phi_{\mathrm{T},-\nu}(x)(q) = \frac{1}{\phi_{\mathrm{T},\nu}(x)(q)}.$$

Indeed triple ratios are computed with respect to flags with the opposite cyclical order, and edge ratios are computed after permuting the second and final arguments.

5.1.2 Duality map in Fock-Goncharov coordinates

Fix a triangulation T and an orientation of S. Two dual structures $x \in \mathcal{T}_3^+(S)$ and $y \in \mathcal{T}_3^-(S)$, are generally represented by different points of $\mathbb{R}_{>0}^{\triangle\cup\underline{E}}$. That is, the composition map

$$\phi_{\mathrm{T}}^- \circ \sigma^+ \circ (\phi_{\mathrm{T}}^+)^{-1} = \phi_{\mathrm{T}}^+ \circ \sigma^- \circ (\phi_{\mathrm{T}}^-)^{-1} \colon \mathbb{R}_{>0}^{\triangle\cup\underline{E}} \to \mathbb{R}_{>0}^{\triangle\cup\underline{E}}$$

is not trivial. Recall from §4.6 that duality maps a flag (V, η) to its dual $(\eta^\perp, V^\perp)$. Referring to the left hand side of Figure 5.1, a straightforward calculation shows that the change of coordinates is locally:

$$e_{20} \mapsto \frac{e_{02}t_{012}(t_{023}+1)}{t_{012}+1}, \qquad e_{02} \mapsto \frac{e_{20}t_{023}(t_{012}+1)}{t_{023}+1} \qquad \text{and} \qquad t_{012} \mapsto \frac{1}{t_{012}}.$$

This transformation manifestly has order two. Its set of fixed points is the algebraic variety $\mathcal{V}$ defined by the polynomials

$$\{e_{ij} - e_{ji},\ t_{ijk} - 1 \mid e_{ij} \in \underline{E},\ t_{ijk} \in \triangle\}.$$

The duality of framings (see also the classification in §5.3) implies that structures in $(\phi_{\mathrm{T}}^{+})^{-1}(\mathcal{V}) \subset \mathcal{T}_3^{+}(S)$ may only have cusps or minimal hyperbolic ends, where the line in the framing of each minimal hyperbolic end must pass through the saddle point. In particular, the space $(\phi_{\mathrm{T}}^{+})^{-1}(\mathcal{V})$ strictly contains the classical Teichmüller space, namely the set of finite-area hyperbolic structures (see Lemma 6.2.1 in §6.2). On the other hand, one can think of the rest of the structures in $(\phi_{\mathrm{T}}^{+})^{-1}(\mathcal{V})$ as infinite-area hyperbolic structures with a preferred framing (and dual framing).

There is a rational re-parametrisation of the edge ratios due to Parreau [42], which reduces the change of coordinates $\phi_{\mathrm{T}}^{-} \circ \sigma^{+} \circ (\phi_{\mathrm{T}}^{+})^{-1}$ into an even simpler form. If e_{ik} is the coordinate of the edge oriented from V_i to V_k, and V_j is the third vertex of the triangle with ordered triple of vertices (V_i, V_j, V_k), then the new edge parameter is

$$s_{ik} := \frac{e_{ik}t_{ijk}}{1 + t_{ijk}}.$$

The duality map on the edge ratios is then reduced to

$$s_{ik} \mapsto s_{ki}.$$

We will not make use of this re-parametrisation of the edge ratios since this will results in a more complicated monodromy map, which is discusses in §5.2.

5.1.3 Transition maps for a different triangulation

The transition map induced by a change of triangulation is slightly more complicated. Henceforth we fix an orientation ν on S and simplify the notation to $\phi_{\mathrm{T}} = \phi_{\mathrm{T},\nu}$. Recall that any two ideal triangulations T and T′ of S differ by a finite sequence of edge flips (cf. Lemma 4.3.2), where a flip along an edge e is the removal of e and insertion of the other diagonal into the arising quadrilateral (see Figure 5.1). In particular, the triangulations T and T′ have the same number of vertices, edges and triangles. Let $\triangle \cup \underline{E}$ and $\triangle' \cup \underline{E}'$ denote triangles and oriented edges of the triangulations T and T′ respectively. Let

$$\Phi_e \colon \mathbb{R}_{>0}^{\triangle \cup \underline{E}} \to \mathbb{R}_{>0}^{\triangle' \cup \underline{E}'}$$

be the coordinate change induced by a flip along e. If $(e_0, \ldots, e_k)$ is the sequence of edges along which we flip in order to get from T to T′, then the coordinate change between T and T′ is the composition map $\Phi_{e_k} \circ \cdots \circ \Phi_{e_0}$. We explicitly give Φ_e below according to Figure 5.1. To simplify the notation, we denote $\phi_{\mathrm{T}}(x)(q), \phi_{\mathrm{T}'}(x)(q)$ by q, q', for all $q \in \triangle \cup \underline{E}$ and $q' \in \triangle' \cup \underline{E}'$.

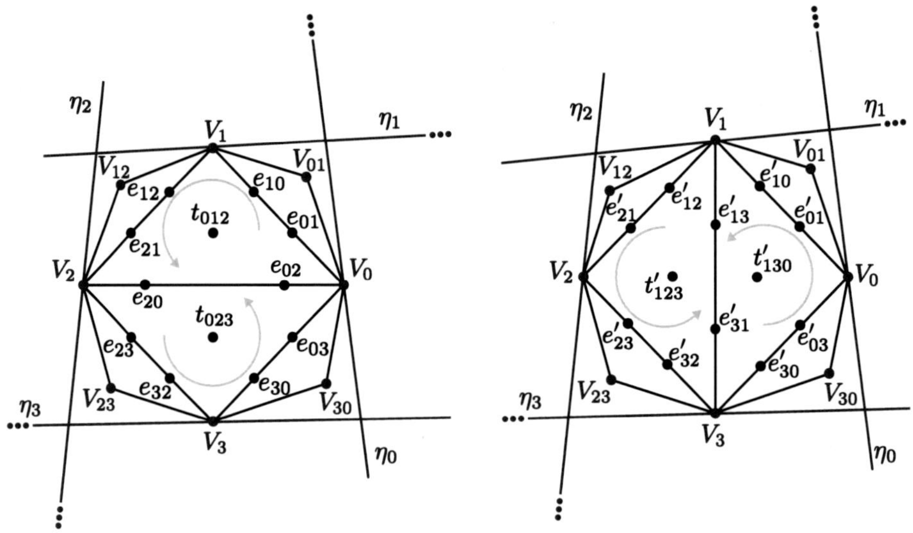

Figure 5.1: The change of coordinates for a flip. The quadrilateral formed by V_i, $i = 0, \ldots, 3$ is assumed to be in an affine patch such that the triangles are oriented anticlockwise.

$$
\begin{aligned}
e'_{01} &= \frac{e_{01}e_{02}}{e_{02}+1}, & e'_{10} &= \frac{e_{10}(e_{02}+1)t_{012}e_{20}}{e_{02}t_{012}e_{20}+t_{012}e_{20}+e_{20}+1},\\
e'_{21} &= e_{21}(e_{20}+1), & e'_{12} &= \frac{e_{12}(e_{02}t_{012}e_{20}+t_{012}e_{20}+e_{20}+1)}{e_{20}+1},\\
e'_{23} &= \frac{e_{23}e_{20}}{e_{20}+1}, & e'_{32} &= \frac{e_{32}e_{02}t_{023}(e_{20}+1)}{e_{02}t_{023}e_{20}+e_{02}t_{023}+e_{02}+1},\\
e'_{03} &= e_{03}(e_{02}+1), & e'_{30} &= \frac{e_{30}(e_{02}t_{023}e_{20}+e_{02}t_{023}+e_{02}+1)}{e_{02}+1},\\
e'_{13} &= \frac{e_{20}+1}{(e_{02}+1)t_{012}e_{20}}, & t'_{130} &= \frac{t_{023}(e_{02}t_{012}e_{20}+t_{012}e_{20}+e_{20}+1)}{e_{02}t_{023}e_{20}+e_{02}t_{023}+e_{02}+1},\\
e'_{31} &= \frac{e_{02}+1}{e_{02}t_{023}(e_{20}+1)}, & t'_{123} &= \frac{t_{012}(e_{02}t_{023}e_{20}+e_{02}t_{023}+e_{02}+1)}{e_{02}t_{012}e_{20}+t_{012}e_{20}+e_{20}+1}.
\end{aligned}
$$

5.2 Monodromy map

In this subsection we develop an efficient way to compute the *monodromy map*, namely the composition $\mu\colon \mathcal{T}_3^+(S) \to \mathfrak{X}^\times(S) \to \mathfrak{X}(S)$. This has an immediate application in §5.3 and plays a key role in the proofs of §6.1 and §6.4.4.

Henceforth, S is assumed to be as in Theorem 4.4.1, endowed with a positively framed convex projective structure $x \in \mathcal{T}_3^+(S)$. We fix an orientation ν and an ideal triangulation T, with triangles and oriented edges $\triangle$ and $\underline{E}$. The canonical isomorphism of Theorem 4.4.1 is given by $\phi = \phi_{T,\nu}$. As earlier, we simplify the notation by reducing $\phi(x)(q), \phi(x)(q')$ to q, q', for all $q \in \triangle \cup \underline{E}$ and $q' \in \triangle' \cup \underline{E}'$. It will always be clear from context whether q stands for a triangle, edge or coordinate.

5.2.1 The monodromy graph

We construct a sub-triangulation T′ of T, by connecting the vertices of every triangle with its barycentre. The *monodromy graph* $\mathcal{G}$ is the dual spine of T′. Let $\widetilde{\mathcal{G}}$ and $\widetilde{\mathrm{T}}$ be lifts of $\mathcal{G}$ and T to Ω, via the developing map of x (Figure 5.2).

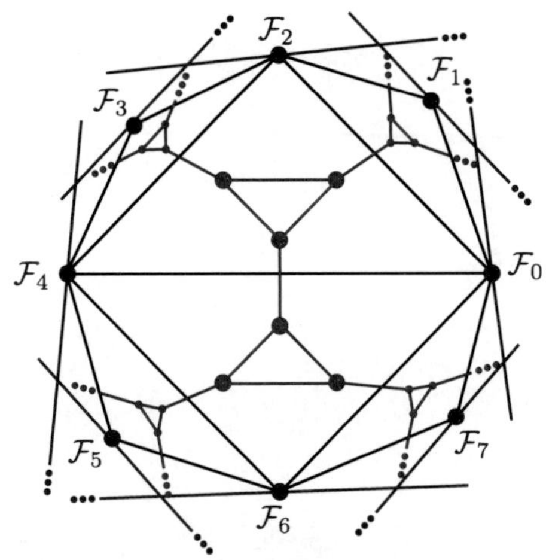

Figure 5.2: The development of the monodromy graph is the dual spine of the development of the triangulation T′.

Let $\underline{E}(\widetilde{\mathcal{G}})$ be the set of oriented edges of $\widetilde{\mathcal{G}}$. We now define the map

$$\mathrm{Mon}\colon\ \underline{E}(\widetilde{\mathcal{G}}) \to \mathrm{SL}(3,\mathbb{R}).$$

Using the construction of Theorem 4.4.1, each triangle t of $\widetilde{\mathrm{T}}$ corresponds to a cyclically ordered triple of flags $\mathcal{F}^t := ((\mathcal{F}^t_i, \mathcal{F}^t_j, \mathcal{F}^t_k))$ in $\mathbb{RP}^2$.

(*i*) Suppose $e \in \underline{E}(\widetilde{\mathcal{G}})$ is strictly contained in t. Then e is an edge of a triangle in $\widetilde{\mathcal{G}}$, which itself is contained in t. It makes sense to compare the orientation of e with that of t. The matrix $\mathrm{Mon}(e)$ is defined as follows.

- If the orientation of e agrees with that of t then

$$\mathrm{Mon}(e)(\mathcal{F}_i, \mathcal{F}_j, \mathcal{F}_k) := (\mathcal{F}_j, \mathcal{F}_k, \mathcal{F}_i).$$

- If the orientation of e does not agree with that of t then

$$\mathrm{Mon}(e)(\mathcal{F}_i, \mathcal{F}_j, \mathcal{F}_k) := (\mathcal{F}_k, \mathcal{F}_i, \mathcal{F}_j).$$

Such an edge is referred to as a $\triangle$*-edge* and is depicted as e_0 in Figure 5.3.

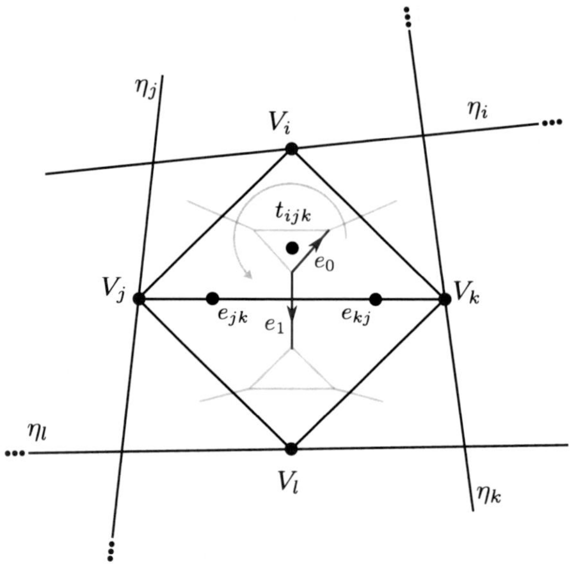

Figure 5.3: If the upper triangle is oriented anticlockwise with respect to the chosen affine patch, then $\mathrm{Mon}(e_0)(\mathcal{F}_i, \mathcal{F}_j, \mathcal{F}_k) = (\mathcal{F}_j, \mathcal{F}_k, \mathcal{F}_i)$ and $\mathrm{Mon}(e_1)(V_i, V_j, V_k, \eta_j\eta_k) = (V_l, V_k, V_j, \eta_j\eta_k)$.

(*ii*) Otherwise, $e \in \underline{E}(\widetilde{\mathcal{G}})$ intersects an edge of $\widetilde{\mathrm{T}}$, say between V_j and V_k. Let t' be the triangle of $\widetilde{\mathrm{T}}$ adjacent to the edge between V_j and V_k, which is different from t. Its associated cyclically ordered triple of flags is $\mathcal{F}^{t'} := (\!(\mathcal{F}_k^{t'}, \mathcal{F}_j^{t'}, \mathcal{F}_l^{t'})\!)$. Suppose e is oriented from t to t'. Then $\mathrm{Mon}(e)$ is the unique projective transformation defined by

$$\mathrm{Mon}(e)(V_i, \mathcal{F}_j, \mathcal{F}_k) = (V_l, \mathcal{F}_k, \mathcal{F}_j).$$

In this case e is called an $\underline{E}$*-edge*. Such an edge is depicted as e_1 in Figure 5.3.

We claim that Mon is well-defined. In case (*i*), such a projective transformation always exists by Theorem 3.7.1 and the fact that $\mathfrak{z}\,(\!(\mathcal{F}_i, \mathcal{F}_j, \mathcal{F}_k)\!) = \mathfrak{z}\,(\!(\mathcal{F}_j, \mathcal{F}_k, \mathcal{F}_i)\!)$. Meanwhile, the operator defined in case (*ii*) can be restated as the unique projective transformation such that

$$\mathrm{Mon}(e)(V_i, V_j, V_k, \eta_j\eta_k) = (V_l, V_k, V_j, \eta_j\eta_k),$$

from which existence and uniqueness are clear by the fact that both quadruples form a projective basis.

Lemma 5.2.1. *Let $\gamma \in \pi_1(S)$. There exists a finite sequence $(e_0, \ldots, e_m)$ of oriented edges $e_i \in \underline{E}(\widetilde{\mathcal{G}})$ such that the path $e_0 \cdot e_1 \cdots e_m \subset \widetilde{\mathcal{G}}$ is a lift of a loop $\widehat{\gamma} \subset \mathcal{G}$,* *freely homotopic to γ. Furthermore,*

$$\mu(x)(\gamma) = \left[\prod_{i=0}^{m} \mathrm{Mon}(e_{m-i})\right],$$

the $\mathrm{SL}(3,\mathbb{R})$ *conjugacy class of* $\prod_{i=0}^{m} \mathrm{Mon}(e_{m-i})$.

Proof. The graph $\mathcal{G}$ contracts in S to the dual spine of T by collapsing all trivial loops. This implies that the natural homomorphism $\pi_1(\mathcal{G}) \to \pi_1(S)$ is surjective.

Let $\gamma \in \pi_1(S)$ and choose a loop in $\mathcal{G}$ which is freely homotopic to γ. Lift this loop to $\widetilde{\mathcal{G}}$ and let $w := e_0 \cdot e_1 \cdots e_m$ be the sequence of edges defined by the chosen lift. Denote by $(\mathcal{F}_0, \mathcal{F}_1, \mathcal{F}_2)$ the flags at the vertices of the triangle t_{012} of $\widetilde{\mathrm{T}}$ from which w starts, and $(\mathcal{F}_3, \mathcal{F}_4, \mathcal{F}_5)$ the flags at the vertices of the triangle t_{345} where w ends. Note that t_{012} and t_{345} are lifts of the same triangle in T as w is a lift of a closed loop.

It is clear from the definition of Mon that $\prod_{i=0}^{m} \mathrm{Mon}(e_{m-i})$ is a projective transformation which maps the vertices and two flags of t_{012} to the vertices and two flags of t_{345}. However $\mathfrak{z}\,(\!(\mathcal{F}_0, \mathcal{F}_1, \mathcal{F}_2)\!) = \mathfrak{z}\,(\!(\mathcal{F}_3, \mathcal{F}_4, \mathcal{F}_5)\!)$ so invoking Theorem 3.7.1,

$$\prod_{i=0}^{m} \mathrm{Mon}(e_{m-i}) \cdot (\mathcal{F}_0, \mathcal{F}_1, \mathcal{F}_2) = (\mathcal{F}_3, \mathcal{F}_4, \mathcal{F}_5).$$

It follows that $\prod_{i=0}^{m} \mathrm{Mon}(e_{m-i})$ agrees on $(\mathcal{F}_0, \mathcal{F}_1, \mathcal{F}_2)$ with a representative in $\mu(x)(\gamma)$. The claim follows from the fact that $\mu(x)(\gamma)$ is (the conjugacy class of) a projective transformation mapping $(\mathcal{F}_0, \mathcal{F}_1, \mathcal{F}_2)$ to $(\mathcal{F}_3, \mathcal{F}_4, \mathcal{F}_5)$ and that such a projective transformation, hence its conjugacy class, is unique. ■

We remark that there was a choice of lift in the above construction. If another lift w' is chosen, giving rise to a product $\prod_{i=0}^{m} \mathrm{Mon}(e'_{m-i})$, and A is the projective transformation taking the starting triangle of w' to the starting triangle of w, then

$$\prod_{i=0}^{m} \mathrm{Mon}(e'_{m-i}) = A^{-1} \cdot \prod_{i=0}^{m} \mathrm{Mon}(e_{m-i}) \cdot A.$$

Hence the product $\prod_{i=0}^{m} \mathrm{Mon}(e_{m-i})$ is defined up to conjugation in $\mathrm{SL}(3,\mathbb{R})$. However, this is the best we can hope for as the monodromy is defined only up to conjugation. Indeed, the domain Ω which is used to define $\widetilde{\mathcal{G}}$, is only defined up to a projective transformation.

Remark 5.2.2. *The sequence in Lemma 5.2.1 can be chosen so that it alternates between $\underline{E}$-edges and $\triangle$-edges. More precisely, we can choose the e_i's so that $(e_0, \ldots, e_m)$ is a sequence of pairs of the form ($\underline{E}$-edge, $\triangle$-edge).*

5.2.2 How to compute the monodromy

Consider the following special case.
Let $\{\mathcal{F}_0, \mathcal{F}_1, \mathcal{F}_2, \mathcal{F}_3\}$, with $\mathcal{F}_i = (V_i, \eta_i)$, be the following generic quadruple of flags:

$$\mathcal{F}_0 = \left(\begin{bmatrix} 0 \\ 0 \\ 1 \end{bmatrix}, [1:0:0] \right), \qquad \mathcal{F}_1 = \left(\begin{bmatrix} 1 \\ -1 \\ 1 \end{bmatrix}, [t_{012} : t_{012}+1 : 1] \right),$$

$$\mathcal{F}_2 = \left(\begin{bmatrix} 1 \\ 0 \\ 0 \end{bmatrix}, [0:0:1] \right), \qquad \mathcal{F}_3 = \left(\begin{bmatrix} e_{02}e_{20} \\ e_{20} \\ 1 \end{bmatrix}, \eta_3 \right).$$

Then $\{V_0, V_1, V_2, \eta_0\eta_2\}$ is a projective basis of $\mathbb{RP}^2$. Moreover, η_1 and V_3 are uniquely determined by the parameters:

$$\not{3}\,(\!(\mathcal{F}_0, \mathcal{F}_1, \mathcal{F}_2)\!) = t_{012},$$

$$\not{4}(\mathcal{F}_0, \mathcal{F}_1, \mathcal{F}_2, \mathcal{F}_3) = e_{02} \qquad \text{and} \qquad \not{4}(\mathcal{F}_2, \mathcal{F}_3, \mathcal{F}_0, \mathcal{F}_1) = e_{20}.$$

Suppose $\{\mathcal{F}_0, \mathcal{F}_1, \mathcal{F}_2, \mathcal{F}_3\}$ are flags assigned to the vertices of some triangulation T, with monodromy graph $\mathcal{G}$. Let e_0 be a $\triangle$–edges in t_{012} and e_1 the $\underline{E}$–edge crossing the interval from V_0 to V_2. Assume e_0 is oriented according to the orientation of t_{012}, and e_1 is oriented towards t_{023}. See Figure 5.3 by setting $(k, i, j, l) = (0, 1, 2, 3)$. Then one easily computes:

$$\mathrm{Mon}(e_0) = \frac{1}{\sqrt[3]{t_{012}}} \begin{pmatrix} 0 & 0 & 1 \\ 0 & -1 & -1 \\ t_{012} & t_{012}+1 & 1 \end{pmatrix},$$

$$\mathrm{Mon}(e_1) = \sqrt[3]{\frac{e_{20}}{e_{02}}} \begin{pmatrix} 0 & 0 & e_{02} \\ 0 & -1 & 0 \\ \frac{1}{e_{20}} & 0 & 0 \end{pmatrix}.$$

If we denote by e_0^{-1}, e_1^{-1} the same edges with opposite orientation, we have that $\mathrm{Mon}(e_0^{-1}) = \mathrm{Mon}(e_0)^{-1}$ and $\mathrm{Mon}(e_1^{-1}) = \mathrm{Mon}(e_1)^{-1}$.

We remark that the appearance of cubic roots can be avoided by replacing triangle invariants and edge invariants by their cubes; this was done in [25].

Lemma 5.2.3. *Let e be an oriented edge of $\widetilde{\mathcal{G}}$ with tail in some triangle t_{ijk} of $\widetilde{\mathrm{T}}$, having flags $\mathcal{F}_i, \mathcal{F}_j$ and $\mathcal{F}_k$ assigned to its vertices. Let $P \in \mathrm{SL}(3, \mathbb{R})$ be the projective transformation mapping $(V_i, V_j, V_k, \eta_k\eta_j)$ to the projective basis $(V_1, V_2, V_0, \eta_0\eta_2)$ as above, and let*

$$T(z) := \frac{1}{\sqrt[3]{z}} \begin{pmatrix} 0 & 0 & 1 \\ 0 & -1 & -1 \\ z & z+1 & 1 \end{pmatrix}, \qquad E(x, y) := \sqrt[3]{\frac{x}{y}} \begin{pmatrix} 0 & 0 & y \\ 0 & -1 & 0 \\ \frac{1}{x} & 0 & 0 \end{pmatrix}.$$

i) If e is a $\triangle$-edge oriented as t_{ijk}, and $\mathfrak{z}\,(\!(\mathcal{F}_i, \mathcal{F}_j, \mathcal{F}_k)\!)= t_{ijk}$, then

$$\mathrm{Mon}(e) = P^{-1} \cdot T(t_{ijk}) \cdot P.$$

ii) If e is an $\underline{E}$-edge crossing the segment of endpoints $(\mathcal{F}_j, \mathcal{F}_k)$ oriented away from t_{ijk}, and e_{kj}, e_{jk} are the corresponding edge ratios, then

$$\mathrm{Mon}(e) = P^{-1} \cdot E(e_{kj}, e_{jk}) \cdot P.$$

Proof. In both cases, right hand side and left hand side agree on the four points $\{V_i, V_j, V_k, \eta_k\eta_j\}$, so the statement follows. ■

We may now compute monodromies of $\mu(x)$ using Theorem 4.4.1.
Let γ be an oriented closed curve on S, in general position with respect to T (and itself). Let $(t_0, \ldots, t_m)$ be the sequence of triangles of $\triangle$ crossed by γ, cyclically ordered with respect to the orientation of γ (possibly with repetitions). For each $i \in \{1, \ldots, m\}$, suppose t_i has the orientation induced by S and that γ enters t_i through the edge $e'_i \in \underline{E}$ and leaves t_i through $e_i \in \underline{E}$, where e_i, e'_i are oriented according to t_i. A diagram of this situation is shown in Figure 5.4. We define the following quantities:

$$E_i := E\left(\phi(x)(e_i), \phi(x)(e_i^{-1})\right), \quad T_i := T\left(\phi(x)(t_i)\right),$$

$$\epsilon_i := \begin{cases} +1 & \text{if } e_i \cap e'_i \text{ is the head of } e'_i, \\ -1 & \text{if } e_i \cap e'_i \text{ is the tail of } e'_i. \end{cases}$$

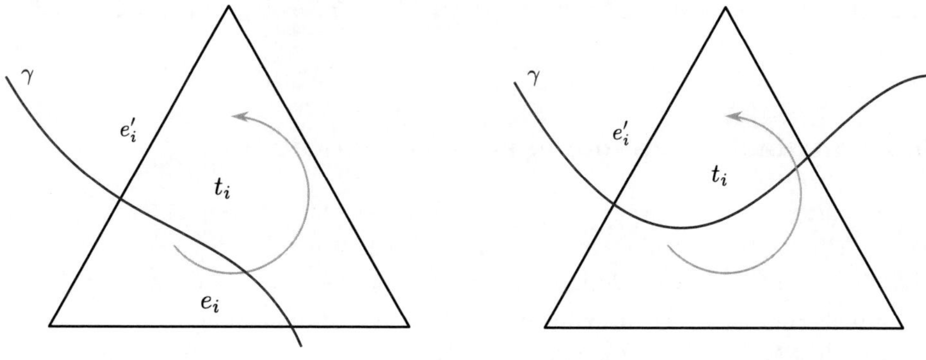

$\epsilon_i = +1$ $\epsilon_i = -1$

Figure 5.4: If we fix an affine patch in which t_i is oriented anticlockwise, $\epsilon_i = +1$ or -1 respectively when γ "turns right" or "turns left".

Theorem 5.2.4. *The monodromy $\mu(x)(\gamma)$ is the conjugacy class $[T_0^{\epsilon_0} \cdot E_0 \cdots T_m^{\epsilon_m} \cdot E_m]$. Furthermore, if γ represents a peripheral element of $\pi_1(S)$ and $\epsilon_0 = 1$ (resp. -1), then $T_0^{\epsilon_0} \cdot E_0 \cdots T_m^{\epsilon_m} \cdot E_m$ is lower triangular (resp. upper triangular).*

Proof. It follows from Lemma 5.2.1, Remark 5.2.2 and Lemma 5.2.3 that $\mu(x)(\gamma)$ is the conjugacy class of

$$Q := \prod_{i=0}^{m} P_{m-i}^{-1} \cdot E_{m-i} \cdot T_{m-i}^{\epsilon_{m-i}} \cdot P_{m-i},$$

where $P_i \in \mathrm{SL}(3, \mathbb{R})$ is the projective transformation corresponding to the triangle t_i, as defined in Lemma 5.2.3, which maps the edge e_i to the oriented edge from V_2 to V_0.
Observe that $P_j \cdot P_{j-1}^{-1} = T_j^{-\epsilon_j} \cdot E_{j-1}^{-1}$ for all $j \in \{1, \ldots, m\}$, as both transformations agree on the four points $\{V_0, V_2, E_{j-1}(V_1), \eta_0\eta_2\}$. Hence

$$\begin{aligned} Q &= \prod_{i=0}^{m} P_{m-i}^{-1} \cdot E_{m-i} \cdot T_{m-i}^{\epsilon_{m-i}} \cdot P_{m-i} \\ &= \left(\prod_{i=0}^{m-1} P_{m-i}^{-1} \cdot E_{m-i} \cdot T_{m-i}^{\epsilon_{m-i}} \cdot T_{m-i}^{-\epsilon_{m-i}} \cdot E_{m-(i+1)}^{-1} \cdot P_{m-(i+1)} \right) \cdot P_0^{-1} \cdot E_0 \cdot T_0^{\epsilon_0} \cdot P_0 \\ &= P_m^{-1} \cdot E_m \cdot T_0^{\epsilon_0} \cdot P_0. \end{aligned}$$

A simple inductive argument shows that $P_m^{-1} = P_0^{-1} \cdot \left(\prod_{i=0}^{m-1} E_i \cdot T_{i+1}^{\epsilon_{i+1}} \right)$. It follows that

$$Q = P_0^{-1} \cdot E_0 \cdot \left(\prod_{i=1}^{m} T_i^{\epsilon_i} \cdot E_i \right) \cdot T_0^{\epsilon_0} \cdot P_0.$$

With regard to the second part of the statement, we remark that γ represents a peripheral element of $\pi_1(S)$ if and only if it circuits some ideal vertex of T. In this case $\epsilon_i = \epsilon_j$ for all $i, j \in \{0, \ldots, m\}$. Since $T_i \cdot E_i$ and $T_i^{-1} \cdot E_i$ are lower- and upper-triangular respectively, the product $T_0^{\epsilon_0} \cdot E_0 \ldots T_m^{\epsilon_m} \cdot E_m$ is triangular as claimed. ■

5.3 Maximal and minimal ends in coordinates

In this paragraph, we show a straightforward application of Theorem 5.2.4. We provide a criterion to determine the geometry of an end using only the Fock-Goncharov coordinates of the structure. More specifically, not only may one distinguish among hyperbolic ends, special ends and cusps, but also between maximality and minimality as well as the different framings. As in the previous sections, S is endowed with a positively framed convex projective structure. We fix an orientation and an ideal triangulation T, and identify $\mathcal{T}_3^+(S)$ with $\mathbb{R}_{>0}^{\triangle \cup E}$ under the canonical isomorphism of Theorem 4.4.1.

Let $\mathcal{E}_i$ be an end of S. Choose a lift $\widetilde{\mathcal{E}_i}$ of $\mathcal{E}_i$ to $\widetilde{S}$. Let V_i be the ideal vertex of $\widetilde{\mathrm{T}}$ in the closure of $\widetilde{\mathcal{E}_i}$ and $\gamma_i \in \pi_1(S)$ an oriented peripheral element whose holonomy fixes V_i. Suppose γ_i is oriented according to the orientation of S. Let $\mathcal{F}_i = (V_i, \eta_i)$ be the framing at V_i. Denote by e_{ik} for $k = 1, \ldots, m_i$ the edges of T oriented away from $\mathcal{E}_i$, and e_{ki} the edges oriented toward $\mathcal{E}_i$. Let t^i_k be the triangle in T appearing after the edge e_{ik} according to the orientation of γ_i (Figure 5.5).

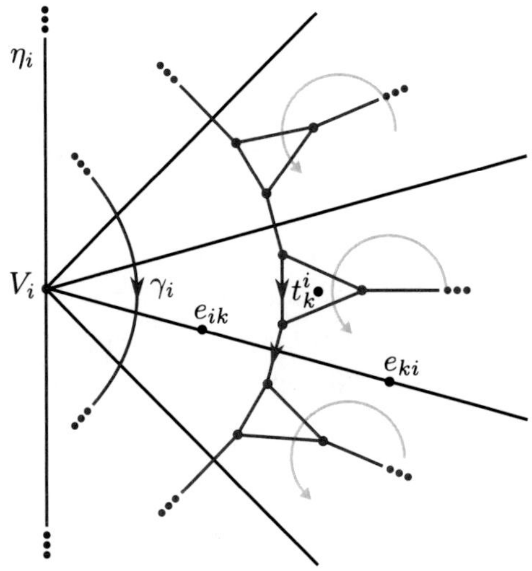

Figure 5.5: The holonomy of the peripheral curve γ_i is a product of pairs $T(t^i_k)E(e_{ik}, e_{ki})$, therefore it is lower triangular.

In Theorem 5.2.4 we show that the conjugacy class of $\mathrm{hol}(\gamma_i)$ has a representative of the form

$$\prod_{k=1}^{m_i} T(t^i_k)E(e_{ik}, e_{ki}) = \prod_{k=1}^{m_i} \sqrt[3]{\frac{e_{ik}}{t^i_k e_{ki}}} \begin{pmatrix} \frac{1}{e_{ik}} & 0 & 0 \\ -\frac{1}{e_{ik}} & 1 & 0 \\ \frac{1}{e_{ik}} & -t^i_k - 1 & t^i_k e_{ki} \end{pmatrix}.$$

The eigenvalues are

$$\lambda_1 = \prod_{k=1}^{m_i} (t^i_k e_{ki} e^2_{ik})^{-\frac{1}{3}} = \mathcal{X}_i^{-\frac{1}{3}} \mathcal{Y}_i^{-\frac{2}{3}}, \qquad \lambda_2 = \prod_{k=1}^{m_i} (t^i_k e_{ki} e^{-1}_{ik})^{\frac{1}{3}} = \mathcal{X}_i^{-\frac{1}{3}} \mathcal{Y}_i^{\frac{1}{3}},$$

$$\lambda_3 = \prod_{k=1}^{m_i} (e^2_{ki} (t^i_k)^2 e_{ik})^{\frac{1}{3}} = \mathcal{X}_i^{\frac{2}{3}} \mathcal{Y}_i^{\frac{1}{3}},$$

where

$$\mathcal{X}_i := \prod_{k=1}^{m_i} t^i_k e_{ki}, \quad \text{and} \quad \mathcal{Y}_i := \prod_{k=1}^{m_i} e_{ik}.$$

As an artefact of the construction in Lemma 5.2.3, the eigenvalue λ_1 always corresponds to the eigenvector V_i, while λ_2 corresponds to another (possibly generalised) eigenvector contained in η_i. Combining this information with the classification of ends gives the following characterisation of the geometry of the ends in terms of the monomials $\mathcal{X}_i$ and $\mathcal{Y}_i$:

1. The end $\mathcal{E}_i$ is hyperbolic:

   (a) The end $\mathcal{E}_i$ is maximal, i.e. V_i is the saddle eigenvector:

   - The line η_i passes through the attracting eigenvector:

   $$\lambda_2 > \lambda_1 > \lambda_3, \quad \text{thus} \quad \mathcal{Y}_i > 1, \text{ and } \mathcal{X}_i\mathcal{Y}_i < 1.$$

   - The line η_i passes through the repelling eigenvector:

   $$\lambda_3 > \lambda_1 > \lambda_2, \quad \text{thus} \quad \mathcal{Y}_i < 1, \text{ and } \mathcal{X}_i\mathcal{Y}_i > 1.$$

   (b) The end $\mathcal{E}_i$ is minimal, i.e. V_i is not the saddle eigenvector:

   - The vertex V_i is the attracting eigenvector and the line η_i passes through the saddle eigenvector:

   $$\lambda_1 > \lambda_2 > \lambda_3, \quad \text{thus} \quad \mathcal{Y}_i < 1, \text{ and } \mathcal{X}_i < 1.$$

   - The vertex V_i is the attracting eigenvector and the line η_i passes through the repelling eigenvector:

   $$\lambda_1 > \lambda_3 > \lambda_2, \quad \text{thus} \quad \mathcal{X}_i\mathcal{Y}_i < 1, \text{ and } \mathcal{X}_i > 1.$$

   - The vertex V_i is the repelling eigenvector and the line η_i passes through the saddle eigenvector:

   $$\lambda_3 > \lambda_2 > \lambda_1, \quad \text{thus} \quad \mathcal{Y}_i > 1, \text{ and } \mathcal{X}_i > 1.$$

   - The vertex V_i is the repelling eigenvector and the line η_i passes through the attracting eigenvector:

   $$\lambda_2 > \lambda_3 > \lambda_1, \quad \text{thus} \quad \mathcal{X}_i\mathcal{Y}_i > 1, \text{ and } \mathcal{X}_i < 1.$$

2. The end $\mathcal{E}_i$ is special:

   (a) The vertex V_i is the attracting eigenvector of multiplicity two:

   - The line η_i does not pass through the repelling eigenvector:

   $$\lambda_1 = \lambda_2 > \lambda_3, \quad \text{thus} \quad \mathcal{Y}_i = 1, \text{ and } \mathcal{X}_i < 1.$$

   - The line η_i passes through the repelling eigenvector:

   $$\lambda_1 = \lambda_3 > \lambda_2, \quad \text{thus} \quad \mathcal{X}_i\mathcal{Y}_i = 1, \text{ and } \mathcal{X}_i > 1.$$

   (b) The vertex V_i is the repelling eigenvector of multiplicity two:

   - The line η_i does not pass through the attracting eigenvector:

   $$\lambda_3 > \lambda_1 = \lambda_2, \quad \text{thus} \quad \mathcal{Y}_i = 1, \text{ and } \mathcal{X}_i > 1.$$

- The line η_i passes through the attracting eigenvector:

$$\lambda_2 > \lambda_1 = \lambda_3, \quad \text{thus} \quad \mathcal{X}_i\mathcal{Y}_i = 1, \text{ and } \mathcal{X}_i < 1.$$

(c) The vertex V_i is the attracting eigenvector of multiplicity one:

$$\lambda_1 > \lambda_2 = \lambda_3, \quad \text{thus} \quad \mathcal{Y}_i < 1, \text{ and } \mathcal{X}_i = 1.$$

(d) The vertex V_i is the repelling eigenvector of multiplicity one:

$$\lambda_2 = \lambda_3 > \lambda_1, \quad \text{thus} \quad \mathcal{Y}_i > 1, \text{ and } \mathcal{X}_i = 1.$$

3. The end $\mathcal{E}_i$ is a cusp:

$$\lambda_1 = \lambda_2 = \lambda_3, \quad \text{thus} \quad \mathcal{Y}_i = 1, \text{ and } \mathcal{X}_i = 1.$$

Chapter 6

Applications of the Parametrisation

6.1 Structures of finite area

We seek now to determine necessary and sufficient conditions for a framed convex projective structure to have finite area, given its representative in $\mathbb{R}_{>0}^{\triangle \cup E}$. This leads to a new proof of a result from Marquis [38].

Throughout this section, the surface $S = S_{g,n}$ will be of negative Euler characteristic, with at least one puncture and orientation ν. Fix a positive framed convex projective structure on S with holonomy representation hol and developing map dev. We denote the holonomy group by Γ. Furthermore, fix an ideal triangulation T of S, let $\Omega := \operatorname{dev}(\widetilde{S}) \subset \mathbb{RP}^2$ and denote by $\widetilde{\mathrm{T}}$ the lift of T to Ω. The map $\phi = \phi_{\mathrm{T},\nu}^{+}$ is the canonical isomorphism defined in Theorem 4.4.1. We identify the simplices $e_{ij}, t_{ijk} \in \mathrm{T}$ with their coordinates $\phi(x)(e_{ij})$ and $\phi(x)(t_{ijk})$ respectively, so as to simplify notation.

Throughout this section, if $\mathcal{C} \subset \mathbb{RP}^2$ is a properly convex domain and $U \subset \mathcal{C}$ then we denote by $\nu_{\mathcal{C}}(U)$ the Hausdorff measure of U with respect to the Hilbert metric on $\mathcal{C}$.

For any properly convex domain $\Omega \subset \mathbb{RP}^2$ and point q, let

$$B_q^{\Omega}(r) = \{x \in \Omega \mid d_{\Omega}(x, q) \leqslant r\}$$

denote the closed metric ball centered at q, of radius r with respect to the Hilbert metric measured on Ω. We will require the following result in order to prove Lemma 6.1.2.

Lemma 6.1.1. *Let $\Omega \subset \mathbb{RP}^2$ be a properly convex domain and let $p \in \Omega$. Then $\nu_{\Omega}(B_p^{\Omega}(r))$ is bounded below by a strictly positive constant depending on r but not on Ω or p.*

Proof. We send Ω to Benzécri position with respect to p, as defined in Theorem 2.3.3. Let $\mathcal{A}$ be the affine patch determined by the construction in Theorem 2.3.3 and just as in that theorem let $B(1)$ and $B(14)$ be the closed balls centred at the origin of radius 1 and 14 respectively with respect to a chosen Euclidean metric on $\mathcal{A}$. In particular, we have

$$B(1) \subset \Omega \subset B(14).$$

Using the fact that the Hilbert metric is order-reversing, this ensures that

$$\nu_\Omega(B_p^\Omega(r)) \geqslant \nu_\Omega(B_p^{B(1)}(r)).$$

However, since $B(1)$ and $B(14)$ are both centered at p, an elementary computation shows that $B_p^{B(1)}(r) = B_p^{B(14)}(\log \frac{15e^r+13}{13e^r+15})$. We now have the following inequality

$$\nu_\Omega(B_p^\Omega(r)) \geqslant \nu_\Omega\left(B_p^{B(14)}\left(\log\frac{15e^r+13}{13e^r+15}\right)\right) \tag{6.1.1}$$

$$\geqslant \nu_{B(14)}\left(B_p^{B(14)}\left(\log\frac{15e^r+13}{13e^r+15}\right)\right). \tag{6.1.2}$$

We have seen in § 2.3.1 that, since $B(14)$ is the interior of a conic, the Hilbert metric on $B(14)$ is twice the hyperbolic metric. The Hausdorff measure, with respect to the hyperbolic metric, of a hyperbolic metric ball of radius R is given by

$$2\pi(\cosh(R) - 1).$$

Therefore the Hausdorff measure with respect to the Hilbert metric, of a Hilbert metric ball of radius R in a conic is given by

$$8\pi\left(\cosh\left(\frac{R}{2}\right) - 1\right).$$

Setting $R = \log\frac{15e^r+13}{13e^r+15}$ we have

$$\nu_{B(14)}\left(B_p^{B(14)}\left(\log\frac{15e^r+13}{13e^r+15}\right)\right) = 8\pi\left(\cosh\left(\frac{1}{2}\log\frac{15e^r+13}{13e^r+15}\right) - 1\right). \tag{6.1.3}$$

Combining equations 6.1.2 and 6.1.3 the result follows. ∎

A proof of the following Lemma was given by Marquis [38]. In order to keep this part of the notes self-contained, we give an adaptation of that proof here.

Lemma 6.1.2. *Let S be a properly convex real projective surface. The area of S is finite if and only if every end of S is a cusp.*

Proof. The surface S has finite area if and only if each ideal triangle in $\widetilde{\mathrm{T}}$ has finite area. Vertices of those ideal triangles correspond to ends of S, therefore it is enough to show that an ideal triangle has finite area if and only if its vertices correspond to cusps.

Fix an ideal triangle t of $\widetilde{\mathrm{T}}$ and denote its vertices by V_0, V_1 and V_2. Let $\mathcal{E}$ be the end corresponding to V_0 and let $\gamma \in \pi_1(S)$ be a generator of the subgroup conjugate to $\pi_1(\mathcal{E})$ such that $\mathrm{hol}(\gamma)$ fixes V_0. Firstly, we show that if $\mathcal{E}$ is not a cusp, the area of t is unbounded. There are two cases, depending on the regularity of $\partial\overline{\Omega}$ at V_0.

Suppose $\mathcal{E}$ is either hyperbolic, or special. If the latter is true further assume that the eigenvalue of V_0 has multiplicity two. It follows from the discussion of the dynamics of peripheral elements that these are the cases where $\partial\overline{\Omega}$ does not have a unique supporting line at V_0. Let η_1 and η_2 be distinct supporting lines to Ω at V_0. One may construct a triangle Δ with vertex V_0, strictly contained in an affine patch $\mathcal{A}$, having two edges on η_1 and η_2, and strictly containing Ω. Observe that Δ and t share the vertex V_0. We are going to show that $\nu_\Delta(t) = \infty$, from which it follows that $\nu_\Omega(t) = \infty$. There is a projective transformation of $\mathbb{RP}^2$ such that

$$V_0 = \begin{bmatrix} 0 \\ 0 \\ 1 \end{bmatrix}, \qquad \text{and} \qquad \Delta = \left\{ \begin{bmatrix} x \\ y \\ 1 \end{bmatrix} \mid x, y > 0 \right\}.$$

In the affine patch $\mathcal{A} : \{z = 1\}$, the vertex V_0 is the origin and Δ is the positive quadrant. Let $\epsilon > 0$ and $0 < a < b$ such that

$$t_0 = \{(x, y) \in \mathcal{A} \mid 0 < x < \epsilon \text{ and } ax < y < bx\} \subset t.$$

Consider the lines $l_a : \{y = ax\}$ and $l_b : \{y = bx\}$. A direct calculation shows that

$$d_\Delta(l_a, P) = \text{constant}, \qquad \text{and} \qquad d_\Delta(V_0, P) = \infty, \quad \forall\, P \in l_b \cap \{0 < x < \epsilon\}.$$

One may therefore construct infinitely many disjoint balls of equal radius which are strictly contained in t_0. It follows from Lemma 6.1.1 that the Hausdorff measure of each such ball is bounded below by a constant depending only on their common radius. Therefore $\nu_\Delta(t_0) = \infty$ and $\nu_\Delta(t) = \infty$.

Now suppose $\mathcal{E}$ is special, but V_0 has an eigenvalue of multiplicity one. We note that the area of S is defined only in terms of Ω and the holonomy group Γ. Therefore we invoke the symmetric group action defined in Theorem 4.5.1 to claim that there exists a framed real projective surface having the same area as S but which is ideally triangulated in such a manner that the eigenvalue associated to the ideal vertex at $\mathcal{E}$ has algebraic multiplicity two. We are now in the case above and the proof follows verbatim.

It remains to show that if each ideal vertex of t corresponds to a cusp, then t has finite area.

Let V_i be an ideal vertex of t. Then the holonomy γ_i of the associated end is a parabolic. This preserves the algebraic horospheres in Ω centred at (η_i, V_i). The parabolic γ_i also preserves a family of ellipsoids passing through V_i and having η_i as a supporting line. This ensures the existence of an open ball B_i, strictly contained in Ω, whose frontier (a conic) shares the supporting line at V_i with Ω. The conic $\partial\overline{B_i}$ intersects $\partial\overline{t}$ in three points, one of which is V_i, and the others will be denoted $t_{i,1}$ and $t_{i,2}$. Let $t_i \subset t$ denote the ideal triangle with vertices V_i, $t_{i,1}$ and $t_{i,2}$. Then t_i is an ideal triangle in B_i, so its Hausdorff measure is 4π and we have the following,

$$4\pi = \nu_{B_i}(t_i) > \nu_\Omega(t_i).$$

The set $C := \overline{t \setminus (t_0 \cup t_1 \cup t_2)} \subset \Omega$ is compact so $\nu_\Omega(C) < \infty$. It follows that

$\nu_\Omega(C \cup t_0 \cup t_1 \cup t_2) < \infty$. Therefore $\nu_\Omega(t) < \infty$. ■

Corollary 6.1.3. *Let S be a properly convex real projective surface. If the area of S is finite, then S is strictly convex.*

Proof. Suppose S has finite area. It follows from Lemma 6.1.2 that every end of S is a cusp. We may assume that each component of the preimage in $\widetilde{S}$ of each end $\mathcal{E}_k$ in S develops into an algebraic horosphere. The compact core S^c of S has finite diameter, D, with respect to the Hilbert metric. In particular, $\Omega = \text{dev}(\widetilde{S})$ is contained in the D–neighbourhood of a countable union of algebraic horospheres.

Now suppose that the structure on S is not strictly convex. Then there is a non-degenerate line segment $s \subset \partial\overline{\Omega}$. It follows from the description of the dynamics of parabolic transformations that if a parabolic fixed point is contained on s, then the whole projective line containing s is contained in $\partial\overline{\Omega}$. This contradicts the fact that Ω is properly convex. Hence no parabolic fixed point is contained in s. Consider a sequence of points $\{p_i\}$ converging to a point in the interior of s. The distance between p_i and any of the algebraic horospheres centred at a parabolic fixed point becomes arbitrarily large as $i \to \infty$. This contradicts the last sentence of the previous paragraph. ■

Theorem 6.1.4 (Marquis 2010, [37]). *The space of finite-area structures $\mathcal{T}_3^f(S)$ is an open cell of dimension $16g - 16 + 6n = -8\chi(S) - 2n$.*

Proof. We have seen in Theorem 4.4.1 that $\mathcal{T}_3^+(S)$ is an open cell of dimension $-8\chi(S) = 16g - 16 + 8n$. It is shown in Lemma 6.1.2 that the condition for an element of $\mathcal{T}_3^+(S)$ to represent a surface of finite area is equivalent to the requirement that every end of S is cuspidal. Fix an ideal vertex V_i of T and cusp neighbourhood ϵ_i of V_i. Let $\gamma_i \in \pi_1(S)$ represent a simple loop such that $\langle\gamma\rangle = \pi_1(\epsilon_i)$. For $k \in \{1, \ldots, m_i\}$ denote by e_{ik} the edges of T oriented away from V_i and by e_{ki} the edges oriented toward V_i. Choose the labelling to agree with the orientation on γ_i. Let t_k^i be the ideal triangle in T incident with the edges e_{ik} and $e_{i(k+1)}$, where the subscripts are read modulo m_i. We understand from the discussion in §5.3, that the holonomy $\text{hol}(\gamma)$ is parabolic if and only if

$$\prod_{k=1}^{m_i} t_k^i e_{ki} =: \mathcal{X}_i = 1, \quad \text{and} \quad \prod_{k=1}^{m_i} e_{ik} =: \mathcal{Y}_i = 1. \tag{6.1.4}$$

As conjugation preserves eigenvalues, the parabolicity conditions reduce to two such equations for each of the n ideal vertices in T. Hereafter, just as we identify $\mathbb{R}_{>0}^{\Delta\cup\underline{E}}$ with $\mathcal{T}_3^+(S)$, we will identify $\mathcal{T}_3^f(S)$ with the subset of $\mathbb{R}_{>0}^{\Delta\cup\underline{E}}$ satisfying equations 6.1.4 without further comment. Therefore $\mathcal{T}_3^f(S)$ is an algebraic subvariety of $\mathcal{T}_3^+(S)$ defined by $2n$ monomial equations. We first show that $\mathcal{T}_3^f(S)$ is smooth and has dimension $-8\chi(S) - 2n$.

For convenience, in what follows we study $\overline{\mathcal{X}}_i := \log(\mathcal{X}_i)$ and $\overline{\mathcal{Y}}_i := \log(\mathcal{Y}_i)$ in place of $\mathcal{X}_i$ and $\mathcal{Y}_i$ respectively, as this will not effect the result and linearise computations. Fix an order on the elements of the canonical basis of $\mathbb{R}_{>0}^{\triangle\cup\underline{E}}$ so that the duals to ideal edges are denoted $e_1, \ldots, e_{-6\chi(S)}$ and duals to ideal triangles are denoted $t_1 \ldots, t_{-2\chi(S)}$. We may now define

$$f : \mathbb{R}_{>0}^{\triangle\cup\underline{E}} \to \mathbb{R}^{2n}$$
$$(e_1, \ldots, e_{-6\chi(S)}, t_1, \ldots, t_{-2\chi(S)}) \mapsto (\overline{\mathcal{X}}_1, \overline{\mathcal{Y}}_1, \ldots, \overline{\mathcal{X}}_n, \overline{\mathcal{Y}}_n).$$

Fix $z \in \mathbb{R}_{>0}^{\triangle\cup\underline{E}}$. We begin by showing that the rows of the Jacobian $J_f(z)$ of f at z are linearly independent, and therefore $J_f(z)$ has rank $2n$. If $1 \leqslant i < 2n$ is odd, we let $\vec{\mathcal{X}}_i(z)$ denote the ith row of $J_f(z)$. If $1 < i \leqslant 2n$ is even, we let $\vec{\mathcal{Y}}_i(z)$ denote the ith row of $J_f(z)$. Suppose that

$$\alpha_1 \vec{\mathcal{X}}_1(z) + \beta_1 \vec{\mathcal{Y}}_1(z) + \cdots + \alpha_n \vec{\mathcal{X}}_n(z) + \beta_n \vec{\mathcal{Y}}_n(z) = \vec{0},$$

for some $\alpha_i, \beta_i \in \mathbb{R}$. Note that

$$\partial_{e_i}\overline{\mathcal{X}}_j(z) = \begin{cases} \frac{1}{z(e_i)} & \text{if } V_j \text{ is the tail of } e_i, \\ 0 & \text{otherwise,} \end{cases} \qquad \partial_{e_i}\overline{\mathcal{Y}}_j(z) = \begin{cases} \frac{1}{z(e_i)} & \text{if } V_j \text{ is the tip of } e_i, \\ 0 & \text{otherwise.} \end{cases}$$

Therefore, for $1 \leqslant i \leqslant -6\chi(S)$ the ith column of $J_f(z)$ has exactly two non-zero entries, one of which is in an even-numbered row, say row j, and one of which is in an odd-numbered row, say row k. In this case $\beta_j = -\alpha_k$. Repeating this for the first $-6\chi(S)$ columns of $J_f(z)$ we see that for all j and k we have $\alpha_j = \alpha_k = -\beta_k = -\beta_j$. Now note that $\partial_{t_i}\overline{\mathcal{Y}}_k(z) = 0$ for all i and k, whereas

$$\partial_{t_i}\overline{\mathcal{X}}_k(z) = \begin{cases} \frac{1}{z(t_i)} & \text{if } V_k \text{ is an ideal vertex of } t_i, \\ 0 & \text{otherwise.} \end{cases}$$

This shows that for $i > -6\chi(S)$, the ith column of $J_f(z)$ has exactly three non-zero entries. Moreover each of these entries appears in an even-numbered row of $J_f(z)$. Let these rows be numbered m, p and q. The ith entry of the vector

$$\alpha_1 \vec{\mathcal{X}}_1(z) + \beta_1 \vec{\mathcal{Y}}_1(z) + \cdots + \alpha_n \vec{\mathcal{X}}_n(z) + \beta_n \vec{\mathcal{Y}}_n(z)$$

must therefore be

$$\frac{\alpha_m + \alpha_p + \alpha_q}{z(t_i)}.$$

However, we have shown above that $\alpha_m = \alpha_p = \alpha_q$. Therefore $\alpha_m = \alpha_p = \alpha_q = 0$. It follows from the above discussion that $\alpha_i = \beta_i = 0$ for all i. Therefore $J_f(z)$ has full rank $2n$ on all of $\mathbb{R}_{>0}^{\triangle\cup\underline{E}}$.

Fix $z \in \mathcal{T}_3^f(S)$. We have shown that the gradient vectors $\vec{\mathcal{X}}_i(z)$ and $\vec{\mathcal{Y}}_j(z)$ for $1 \leqslant i, j \leqslant n$ are linearly independent in $T_z\mathbb{R}_{>0}^{\triangle\cup\underline{E}}$. The tangent space $T_z\mathcal{T}_3^f(S)$ is the subspace of $T_z\mathbb{R}_{>0}^{\triangle\cup\underline{E}}$ containing the directions along which $\overline{\mathcal{X}}_i$ and $\overline{\mathcal{Y}}_i$ are constant for all i. This is the intersection of the $2n$ co-dimension–1 subspaces of $T_z\mathbb{R}_{>0}^{\triangle\cup\underline{E}}$ which are perpendicular to both $\vec{\mathcal{X}}_i(z)$ and $\vec{\mathcal{Y}}_j(z)$ for all $i, j = 1, \ldots, n$. Since $\vec{\mathcal{X}}_i(z)$ and $\vec{\mathcal{Y}}_j(z)$ are linearly independent, the implicit function theorem implies that this subspace has co-dimension–$2n$ in $T_z\mathbb{R}_{>0}^{\triangle\cup\underline{E}}$. This is independent of the choice of $z \in \mathcal{T}_3^f(S)$, thus it follows that $\mathcal{T}_3^f(S)$ is a smooth subvariety of dimension $-8\chi(S) - 2n$ as required.

It remains to show that $\mathcal{T}_3^f(S)$ is a topological cell. Having seen that $\mathcal{T}_3^f(S)$ is a smooth subvariety of the required dimension, we now use equations 6.1.4 to express $2n$ of the coordinate functions $e_1, \ldots, e_{-6\chi(S)}, t_1, \ldots, t_{-2\chi(S)}$ as a monomial function in the remaining $-8\chi(S) - 2n$ coordinate functions. Indeed the only obstruction to doing so would be if one of the equations 6.1.4 were obtained as a consequence of the others. However, the fact that $\mathcal{T}_3^f(S)$ has dimension $-8\chi(S) - 2n$ ensures that this is not the case. Up to reordering coordinate functions we may assume that

$$\begin{aligned} e_i &= \phi_i(e_{x+1}, \ldots, e_{-8\chi(S)}, t_{y+1}, \ldots, t_{-2\chi(S)}), \quad \text{for } 1 \leqslant i \leqslant x, \text{ and} \\ t_k &= \psi_k(e_{x+1}, \ldots, e_{-8\chi(S)}, t_{y+1}, \ldots, t_{-2\chi(S)}), \quad \text{for } 1 \leqslant k \leqslant y, \end{aligned}$$

where ϕ_i and ψ_k are monomials and $x + y = 2n$. We now have a well-defined projection map

$$\pi : \mathcal{T}_3^f(S) \to \mathbb{R}_{>0}^{-8\chi(S)-2n}$$

which we obtain by forgetting e_i and t_k for $1 \leqslant i \leqslant x$, and $1 \leqslant k \leqslant y$. This map clearly has a continuous inverse obtained by setting e_i to take the same value as the monomial ϕ_i for $1 \leqslant i \leqslant x$ and t_k to have the same value as ψ_k for $1 \leqslant k \leqslant y$. We can extend this definition of π^{-1} to all of $\mathbb{R}_{>0}^{-8\chi(S)-2n}$. To show that π is surjective we need to ensure that the pre-image thus obtained has strictly positive coordinate functions. This is so because the functions ϕ_i and ψ_k are monomials with positive coefficients and are therefore positive definite on $\mathbb{R}_{>0}^{-8\chi(S)-2n}$. Therefore π is a homeomorphism as required. ■

6.2 The classical Teichmüller space

Having obtained a subvariety of $\mathcal{T}_3^+(S)$ comprising structures of finite area there is another obvious subvariety that we wish to interrogate – the set of points which are invariant under the projective duality $\phi_{\mathrm{T}}^- \circ \sigma^+ \circ (\phi_{\mathrm{T}}^+)^{-1}$ defined in §5.1.2. This is of interest because, for instance, results dating back to Kay [31] indicate that these structures should be identified with finite area hyperbolic structures on S. To that end we use this subsection to characterise the Fock-Goncharov coordinates of these structures and in so doing reproduce a classical result of Teichmüller [47] regarding the topology of the Teichmüller space Teich(S) of S.

As usual $S = S_{g,n}$ is a surface of negative Euler characteristic, with at least one puncture and orientation ν. We fix an ideal triangulation T of S, and let $\phi = \phi_{\mathrm{T},\nu}^+$ be the canonical isomorphism constructed in Theorem 4.4.1.

A quick calculation shows that the Fock-Goncharov coordinates which are invariant under projective duality are

$$\begin{aligned} \phi(x)(e) &= \phi(x)(e^{-1}), && \forall\, e \in \underline{E}, \\ \phi(x)(t) &= 1, && \forall\, t \in \triangle. \end{aligned} \tag{6.2.1}$$

One can easily deduce from the classification in §5.3 that this can only happen when each end is either a cusp or a minimal hyperbolic end, framed by the line through the saddle point. If we restrict our attention to the former, the following result elucidates how we identify the convex projective structure in question with a finite area hyperbolic structure.

Lemma 6.2.1. *Let* $x = (\Omega, \Gamma, f)^+ \in \mathcal{T}_3^+(S)$. *For every ideal vertex* V_i *of* T, *let* $e_{ik} \in \underline{E}$, $k = 1, \ldots, m_i$, *be the oriented edges starting from* V_i. *Then* $\partial\overline{\Omega}$ *is a conic of* $\mathbb{RP}^2$ *if and only if*

$$\begin{aligned} \phi(x)(e) &= \phi(x)(e^{-1}), && \forall\, e \in \underline{E}, \\ \phi(x)(t) &= 1, && \forall\, t \in \triangle, \\ \prod_{k=1}^{m_i} \phi(e_{ik}) &= 1, && \forall\, i = 1, \ldots, n. \end{aligned} \tag{6.2.2}$$

Proof. We begin with a preliminary observation. Suppose $\{\mathcal{F}_0, \mathcal{F}_1, \mathcal{F}_2, \mathcal{F}_3\}$ is a generic quadruple of flags of $\mathbb{RP}^2$, where $\mathcal{F}_i = (V_i, \eta_i)$. Let $\mathcal{C}$ be the unique conic passing through V_0, V_1, V_2 and tangent to η_0 and η_1. A straightforward computation shows that $\mathcal{C}$ is tangent to η_2 if and only if $\not{3}\,(\!(\mathcal{F}_0, \mathcal{F}_1, \mathcal{F}_2)\!)= 1$ (see § 3.6). Furthermore, $V_3 \in \mathcal{C}$ if and only if $\not{4}(\mathcal{F}_0, \mathcal{F}_1, \mathcal{F}_2, \mathcal{F}_3) = \not{4}(\mathcal{F}_2, \mathcal{F}_3, \mathcal{F}_0, \mathcal{F}_1)$.

Let $\widetilde{\mathrm{T}}$ be the lift of T to Ω. By Lemma 6.1.2, the projective structure x has finite volume if and only if all ends are cusps. In that case, the set of vertices of $\widetilde{\mathrm{T}}$ is dense in $\partial\overline{\Omega}$, therefore $\partial\overline{\Omega}$ is a conic if and only if there is a conic passing through all the vertices of $\widetilde{\mathrm{T}}$. We can conclude that $\partial\overline{\Omega}$ is a conic if and only if

a) for each vertex V_i,

$$\prod_{k=1}^{m_i} t^i_k e_{ki} = 1 \quad \text{and} \quad \prod_{k=1}^{m_i} e_{ik} = 1, \tag{6.2.3}$$

where $\{t^i_k\}_{k=1,\ldots,m_i}$ is the set of triangles in T with a vertex at V_i. These conditions force the ends to be cusps, as underlined in the classification in §5.3.

b) for all edges $e \in \underline{E}$ and triangles $t \in \triangle$,

$$\phi(x)(e) = \phi(x)(e^{-1}) \quad \text{and} \quad \phi(x)(t) = 1. \tag{6.2.4}$$

This follows from the preliminary discussion. In fact, assuming all ends are cusps, one can construct a conic through all the vertices of T, tangent to the supporting lines defined by the flags, if and only if every triple ratio is equal to 1 and edge ratios of opposite oriented edges are equal.

Combining equations (6.2.3) and (6.2.4) gives precisely the conditions of the lemma. ■

The Klein disc embeds conformally in $\mathbb{RP}^2$ as the interior of a conic and the group of orientation-preserving hyperbolic isometries, $\mathrm{PSL}(2,\mathbb{R})$, is isomorphic to $PO^+(1,2) \leqslant \mathrm{SL}(3,\mathbb{R})$. Hence every hyperbolic structure on S is naturally a convex projective structure on S. Conversely if Ω is a conic then, up to conjugation, the corresponding holonomy group is a subgroup of $\mathrm{Aut}(\Omega) \cong PO^+(1,2)$. Moreover the Hilbert metric on a conic is twice the natural hyperbolic metric in the Klein model. Two hyperbolic structures are equivalent as hyperbolic structures if and only if they are equivalent as projective structures.

We have seen that finite-area structures admit a unique natural framing (resp. dual framing) so one can canonically identify Teich(S) with a subvariety of $\mathcal{T}_3^+(S)$ despite the fact that Teich(S) contains no information regarding a framing. Thus we deduce the following result.

Theorem 6.2.2 (Teichmüller 1939, [47]). *The Teichmüller space of $S = S_{g,n}$ is homeomorphic to $\mathbb{R}^{6g-6+2n}$.*

Proof. The parameter space Teich(S) is the set of those structures $(\Omega, \Gamma, f)^+ \in \mathcal{T}_3^+(S)$ such that $\partial\overline{\Omega}$ is a conic of $\mathbb{RP}^2$. Recall from Theorem 4.4.1 that $\mathcal{T}_3^+(S)$ is an open cell of dimension $-8\chi(S) = 16g - 16 + 8n$. By Lemma 6.2.1, Teich$(S)$ is a subvariety of $\mathcal{T}_3^+(S)$ defined by the $(-5\chi(S) + n)$ equations in (6.2.2). This set of equations is independent, thus

$$\dim_{\mathbb{R}}(\mathrm{Teich}(S)) = -3\chi(S) - n = 6g - 6 + 3n - n = 6g - 6 + 2n.$$

Furthermore, there is a natural retraction of $\mathcal{T}_3^+(S)$ onto Teich(S), thus Teich(S) is also an open cell, and the claim follows. ■

6.3 Closed surfaces

A result of Goldman [21] determines that if S is an orientable, closed surface then $\mathcal{T}_3(S)$ is an open cell of dimension $-8\chi(S)$. We are going to give a different proof of the same result, using Fock-Goncharov coordinates. This requires a notion of how to glue two ends of a convex projective surface so as to induce a convex projective structure on the resulting glued surface.

Bonahon and Dreyer [3] also use Fock and Goncharov's coordinates to prove this result employing geodesic laminations in place of ideal triangulations. We present this version here to motivate the usefulness of having a simple expression for the monodromy map.

Theorem 6.3.1 (Goldman 1990, [21]). *If S is a closed, orientable surface of negative Euler characteristic, then $\mathcal{T}_3(S)$ is an open cell of dimension $-8\chi(S)$.*

Proof. Suppose S is of genus g and is endowed with a fixed convex projective structure (Ω, Γ, f). That is, S is the quotient of a properly convex domain $\Omega \subset \mathbb{RP}^2$ by the action of a discrete subgroup $\Gamma < \mathrm{SL}(3,\mathbb{R})$ marked by f. Denote the holonomy representation of S by hol and developing map by dev.

Choose an essential, simple, closed, non-separating geodesic with respect to the Hilbert metric on S and denote it by γ. Let F be the interior of the surface obtained from S by cutting along γ, and let f_0 be the restriction of f to F. Then F is a topologically finite surface, of the same Euler characteristic as S, with two ends, say $\mathcal{E}_+$ and $\mathcal{E}_-$. Let $\gamma_\pm$ be the generator of the peripheral subgroup $\pi_1(\mathcal{E}_\pm)$, homotopic to γ in S. The restriction of hol to curves in F endows F with a holonomy representation which we denote by hol_0. Moreover, F inherits a developing map from S, as the universal cover $\widetilde{S}$ of S is tiled by copies of the universal cover of F. Choose one such copy of the universal cover of F and name it $\widetilde{F}$. Let dev_0 be the developing map obtained by restricting dev to $\widetilde{F}$. It is clear that $\Omega_0 := \mathrm{dev}_0(\widetilde{F}) \subset \mathbb{RP}^2$ is a convex domain preserved by the action of $\Gamma_0 := \mathrm{hol}_0(\pi_1(F))$, thus $(\Omega_0, \Gamma_0, f_0)$ is a convex projective structure on F.

The surface F has minimal hyperbolic ends at both $\mathcal{E}_+$ and $\mathcal{E}_-$ as the peripheral elements at those ends are inherited from the geometry of S. We frame $\mathcal{E}_+$ and $\mathcal{E}_-$ by choosing the Γ_0–orbit of the attracting fixed point and the line through the saddle point. This choice is arbitrary, but it gives a well-defined map

$$\Psi : \mathcal{T}_3(S) \to \mathcal{T}_3^+(F).$$

By construction, $\Psi(\mathcal{T}_3(S))$ is a subset of the set of structures $\mathcal{V} \subset \mathcal{T}_3^+(F)$, wherein:

(*a*) All ends are minimal hyperbolic ends, framed by the attracting fixed points and the lines through the saddle points,

(*b*) The transformation $\mathrm{hol}_0(\gamma_+)$ is conjugate to $\mathrm{hol}_0(\gamma_-)$.

From the classification in §5.3, we know that the set of structures satisfying (*a*) is an open cell of $\mathcal{T}_3^+(F)$, hence of full dimension. On the other hand, condition (*b*) is equivalent to requiring that the eigenvalues of $\mathrm{hol}_0(\gamma_+)$ and $\mathrm{hol}_0(\gamma_-)$ are equal. This amounts to two equations easily deducible from the eigenvalues of a peripheral element which are calculated in §5.3. Therefore $\mathcal{V}$ is an open cell of dimension $-8\chi(F) - 2$.

We are going to show that $\Psi(\mathcal{T}_3(S)) = \mathcal{V}$. Let $\gamma_\pm$ be the generator of $\pi(\mathcal{E}_\pm)$ such that the attracting fixed point of $\mathrm{hol}_0(\gamma_\pm)$ is one of the vertices of the framing of $\mathcal{E}_\pm$. By definition, $\mathrm{hol}_0(\gamma_+)$ and $\mathrm{hol}_0(\gamma_-)$ are conjugate, so let $A \in \mathrm{SL}(3,\mathbb{R})$ be a conjugating matrix

$$\mathrm{hol}_0(\gamma_+) = A^{-1} \cdot \mathrm{hol}_0(\gamma_-) \cdot A$$

with the property that A maps the attracting (resp. repelling, saddle) fixed point of $\mathrm{hol}_0(\gamma_+)$ to the attracting (resp. repelling, saddle) fixed point of $\mathrm{hol}_0(\gamma_-)$. This is done with respect to a lift of Ω to $\mathbb{S}^2$.

Define the group $\Gamma := \langle \Gamma_0, A \rangle < \mathrm{SL}(3,\mathbb{R})$. Let s be the open segment between the attracting and repelling fixed points of $\mathrm{hol}_0(\gamma_-)$. Hence define Ω to be the orbit of $\Omega_0 \cup s$ under Γ.

We claim that the domain Ω is properly convex, tessellated by copies of Ω_0, and that $\Omega/\Gamma \cong S$. By definition, Ω_0 and $A \cdot \Omega_0$ share supporting projective lines at the endpoints of s. This ensures that $A \cdot \Omega_0$ is contained in the interior of the closed triangle δ defined by the fixed points of $\mathrm{hol}_0(\gamma_-)$ which contains the interval s and which does not contain Ω_0. Both Ω_0 and $A \cdot \Omega_0$ are convex, and so $\Omega_0 \cup A \cdot \Omega_0$ is convex. Iterating this argument at all lifts of $\mathcal{E}_+$ and $\mathcal{E}_-$ shows that Ω is convex and tesselated by copies of Ω_0. This also implies that $\Omega/\Gamma \cong S$.

Let f be the unique homeomorphism extending f_0 to S. Then (Ω, Γ, f) is a convex projective structure of S, which maps onto $(\Omega_0, \Gamma_0, f_0)$ via Ψ. It follows that Ψ surjects onto $\mathcal{V}$.

To conclude the proof, it is sufficient to show that the fibers of the map Ψ are 2–dimensional discs. This follows from the fact that the centraliser of $\mathrm{hol}_0(\gamma_\pm)$ in $\mathrm{SL}(3, \mathbb{R})$ is a 2–dimensional subgroup. Since A is conjugate to a diagonal matrix with positive eigenvalues, this implies there is a 2–disc worth of choices for A. ■

In fact, more is known about the moduli space for a closed surface. Recall from §2.2.3 that $\mathfrak{X}_3(S) = \mathrm{Hom}(\pi_1(S), \mathrm{SL}(3, \mathbb{R})) /\!\!/ \mathrm{SL}(3, \mathbb{R})$ is the character variety of S. In the case of a closed surface, the monodromy map $\mu\colon \mathcal{T}_3(S) \to \mathfrak{X}_3(S)$ is known to give a Zariski component:

Theorem 6.3.2 (Choi-Goldman 1997, [8]). *If S is a closed, orientable surface of negative Euler characteristic, then $\mathcal{T}_3(S)$ is homeomorphic to a connected component of $\mathfrak{X}_3(S)$.*

6.4 A Poisson structure on the moduli space

This section is devoted to the construction and analysis of a Poisson structure on the moduli space $\mathcal{T}_3^+(S_{g,n})$. We start with a brief summary of basic notions on symplectic structures and Poisson structures. In §6.4.2 we remind the reader about Goldman's results on character varieties, with particular emphasis on Goldman's formula (Theorem 6.4.5). Next, we shift our attention to the moduli space $\mathcal{T}_3^+(S_{g,n})$, and explicitly construct a Poisson bracket via Theorem 4.4.1 and Fock-Goncharov moduli space. Finally, we concentrate on the relation between the Poisson structures on $\mathcal{T}_3^+(S_{g,n})$ and on $\mathfrak{X}_3(S_{g,n})$. The material in §6.4.4 is devoted to developing the necessary tools to show that the structures are homomorphic.

6.4.1 Symplectic structures and Poisson structures

Over Riemannian manifolds, a smooth scalar function f differentiates to a vector field ∇f via the metric. Symplectic manifolds are similarly rich in structure, they relate scalar functions to Hamiltonian vector fields.

A *symplectic form* ω on a smooth manifold M is a closed differential 2–form which is not degenerate in that the kernel of the operation of contracting vector fields is trivial. The pair (M, ω) is a *symplectic manifold.* Because ω is non-degenerate, each smooth scalar function f corresponds to a unique *Hamiltonian vector field* X_f, the vector field satisfying

$$\omega(X_f(x), v_x) = -D_x f(v_x), \qquad \text{for all } x \in M \text{ and } v_x \in T_x M.$$

A *Poisson manifold* is a smooth manifold M endowed with a *Poisson bracket, i.e.* a Lie bracket

$$\{-,-\} : C^\infty(M) \times C^\infty(M) \to C^\infty(M)$$

on the algebra $C^\infty(M)$ of smooth functions, satisfying the Leibniz Rule.

Lemma 6.4.1. *Let (M, ω) be a symplectic manifold. For all $f, g \in C^\infty(M)$, the bracket $\{f, g\}_\omega := \omega(X_f, X_g)$ is a Poisson bracket.*

Symplectic manifolds are Poisson manifolds, but the converse is not true. A classical example of a Poisson manifold which may not give rise to a compatible symplectic structure is the linear Poisson structure on the dual of a Lie algebra. The fact that a symplectic form must be non-degenerate ensures that a symplectic manifold must have even dimension which clearly does not have to be the case for a Lie algebra.

If the Poisson bracket of a Poisson manifold M restricts to a submanifold $N \subset M$, then N is said to be a *Poisson sub-manifold.* If moreover the restricted Poisson bracket on N can be defined by a symplectic form, then N is a *symplectic submanifold.* Any Poisson manifold admits a *symplectic foliation*, a possibly singular foliation whose leaves are the maximal symplectic submanifolds.

6.4.2 Goldman's Poisson structure and formula

Let $S_{g,n}$ be a compact oriented surface with fundamental group π_1. Let G be a Lie group with an $\mathrm{Ad}(G)$–invariant non-degenerate bilinear form on its Lie algebra. The *G–character variety* $\mathfrak{X}_G := \mathrm{Hom}(\pi_1, G) /\!\!/ G$ is the categorical quotient of the G–representation variety $\mathrm{Hom}(\pi_1, G)$ by the action of G under conjugation. We begin by recalling the following:

Theorem 6.4.2 (Goldman 1984, [17]). *If $S_{g,n}$ is closed ($n = 0$), then $\mathfrak{X}_G$ has a natural symplectic structure.*

Theorem 6.4.2 is valid for any reductive group G, in particular $\mathrm{SL}(m, \mathbb{R})$. Goldman's result was generalised to compact oriented surfaces with non-empty boundary in the following sense.

Theorem 6.4.3 (Guruprasad-Huebschmann-Jeffrey-Weinstein 1997, [24]). *If $S_{g,n}$ has $n > 0$ boundary components, then $\mathfrak{X}_G$ has a natural Poisson structure.*

On each leaf of its symplectic foliation, the symplectic form is that defined by Goldman. For the purpose of describing this foliation, we fix a presentation of π_1

based at $p \in S_{g,n}$:

$$\pi_1 := \pi_1(S_{g,n}, p) = \langle x_1, y_1, \dots, x_g, y_g, b_1, \dots, b_n \mid \prod_{i=1}^{g}[x_i, y_i]\prod_{j=1}^{n} b_j = 1\rangle.$$

For $n > 0$, the fundamental group π_1 is free of rank $r = 2g+n-1 = 1-\chi(S_{g,n})$. The *i–th boundary map* $\mathbf{b_i} : [\rho] \mapsto [\rho_{|b_i}]$ sends the class of a representation to the class corresponding to the restriction of ρ to the boundary loop b_i. The *boundary map*

$$\mathbf{b} = (\mathbf{b_1}, \dots, \mathbf{b_n}) : \mathfrak{X}_G \to (G /\!\!/ G)^{\times n}$$

is a submersion on a Zariski-open subset $\mathfrak{X}_G^{\mathbf{b}} \subset \mathfrak{X}_G$ (cf. Hartshorne [26], page 271), and $\mathfrak{X}_G^{\mathbf{b}}$ is foliated by non-singular submanifolds $\mathbf{F}_y := \mathfrak{X}^{\mathbf{b}} \cap \mathbf{b}^{-1}(y)$. The leaves $\mathbf{F}_y$ are symplectic submanifolds (cf. Lawton [35]), which extend to make the complement of the singular locus in $\mathfrak{X}_G$ a Poisson manifold. Furthermore, the Poisson structure extends continuously over singularities in $\mathfrak{X}_G$ (cf. Goldman [17]), making it into a *Poisson variety*, an affine variety whose coordinate ring is a Poisson algebra.

As further motivation for investigating structures on the G–character variety, we recall an application of Theorem 6.4.3 to the special case $G = \mathrm{SL}(m, \mathbb{R})$, where the Poisson structure of $\mathfrak{X}_{\mathrm{SL}(m,\mathbb{R})}$ distinguishes the topology of the surface, whereas the algebraic structure alone does not.

Theorem 6.4.4 (Lawton 2009, [35]). *Let S and S' be compact connected orientable surfaces with boundary, and let $\mathfrak{X}_m(S) = \mathfrak{X}_{\mathrm{SL}(m,\mathbb{R})}(S)$ and $\mathfrak{X}_m(S') = \mathfrak{X}_{\mathrm{SL}(m,\mathbb{R})}(S')$ be their respective* SL$(m, \mathbb{R})$*–character varieties. Then:*

(1) *$\mathfrak{X}_m(S)$ and $\mathfrak{X}_m(S')$ are isomorphic as varieties if and only if $\chi(S) = \chi(S')$;*

(2) *$\mathfrak{X}_m(S)$ and $\mathfrak{X}_m(S')$ are isomorphic as Poisson varieties if and only if $S \cong S'$.*

When G is a classical matrix group (for example GL, SL or SU), the Poisson structure of $\mathfrak{X}_G$ has a deep geometric characterisation (see Goldman [17] and Lawton [35]). Henceforth we restrict our attention to the case $G = \mathrm{SL}(m, \mathbb{R})$ which is of main interest here ($m = 3$), and abbreviate the notation of the SL$(m, \mathbb{R})$–character variety to $\mathfrak{X}_m$.

Let $\mathrm{tr} : \mathrm{SL}(m, \mathbb{R}) \to \mathbb{R}$ be the *trace function.* For all $\alpha \in \pi_1$, define

$$\begin{aligned} \mathrm{tr}_\alpha \colon \mathfrak{X}_m &\to \mathbb{R}, \\ [\rho] &\mapsto \mathrm{tr}(\rho(\alpha)). \end{aligned}$$

The subalgebra of $C^\infty(\mathfrak{X}_m)$ generated by these functions is the *trace algebra* $\mathrm{Tr}(\mathfrak{X}_m)$.

Let α and β be immersed curves in S representing elements in π_1 (in general position). Denote the *set of (transverse) double point intersections* of α and β by $\alpha \cap \beta$, and the *oriented intersection number* of α and β at a point $p \in \alpha \cap \beta$ by $\epsilon(p, \alpha, \beta) \in \{-1, 1\}$. Additionally, let $\alpha_p \in \pi_1(S, p)$ represent the curve α based at the point p.

Theorem 6.4.5 (Goldman 1986, [18], Lawton 2009, [35]). *Let $\{-,-\}_{\mathrm{Gol}}$ be the Poisson structure on $\mathfrak{X}_m$ from Theorem 6.4.3. Then $\{-,-\}_{\mathrm{Gol}}$ restricts to a Poisson bracket on* $\mathrm{Tr}(\mathfrak{X}_m)$ *and*

$$\{\mathrm{tr}_\alpha, \mathrm{tr}_\beta\}_{Gol} = \sum_{p\in\alpha\cap\beta} \epsilon(p,\alpha,\beta)(\mathrm{tr}_{\alpha_p\beta_p} - \frac{1}{n}\mathrm{tr}_\alpha \mathrm{tr}_\beta), \tag{6.4.1}$$

It is worth noticing that one may rewrite (6.4.1) as

$$\{\mathrm{tr}_\alpha, \mathrm{tr}_\beta\}_{Gol} = \left(\sum_{p\in\alpha\cap\beta} \epsilon(p,\alpha,\beta)\,\mathrm{tr}_{\alpha_p\beta_p}\right) - \frac{\iota(\alpha,\beta)}{n}\mathrm{tr}_\alpha \mathrm{tr}_\beta, \tag{6.4.2}$$

for the *algebraic intersection number* $\iota(\alpha,\beta)$. However we can not simplify it any further as $\mathrm{tr}_{\alpha_p\beta_p}$ heavily depends on p, as shown in Figure 6.1.

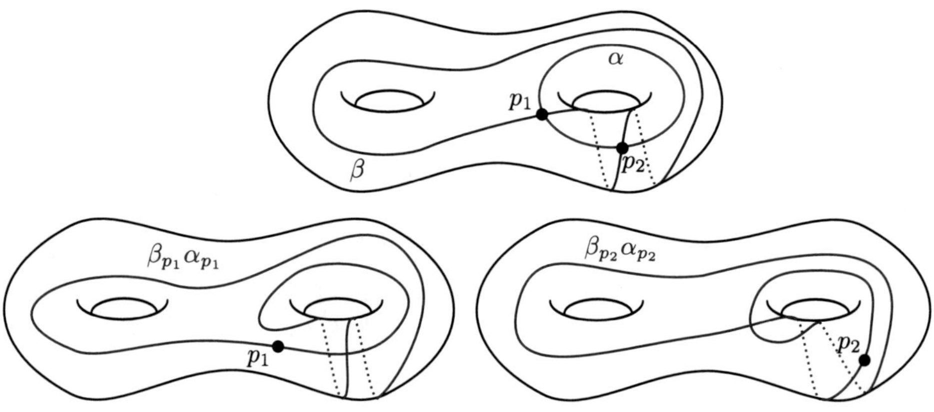

Figure 6.1: The curves $\alpha_{p_1}\beta_{p_1}$ and $\alpha_{p_2}\beta_{p_2}$ are not homotopic.

For complex numbers, it is known that the coordinate ring $\mathbb{C}[\mathfrak{X}_{\mathrm{SL}(m,\mathbb{C})}]$ of the $\mathrm{SL}(m,\mathbb{C})$–character variety is generated by trace functions, hence Theorem 6.4.5 is true for $\mathrm{Tr}(\mathfrak{X}_{\mathrm{SL}(m,\mathbb{C})}) = \mathbb{C}[\mathfrak{X}_{\mathrm{SL}(m,\mathbb{C})}]$ (cf. Procesi [43]). However, it is still an open problem for $\mathrm{SL}(m,\mathbb{R})$.

6.4.3 Fock and Goncharov's Poisson structure

We recall the setting and notation of §4. As usual, T denotes a fixed ideal triangulation of a surface $S = S_{g,n}$ $(2g+n>2)$ and $\triangle, \underline{E}$ are the sets of its triangles and oriented edges respectively. To frame our subsequent definition, we will make the following assumption:

(I) *Every edge in* T *is contained in two distinct triangles, or equivalently every triangle has three distinct edges.*

Such a triangulation always exists, and can be constructed algorithmically from any given triangulation. When t is a triangle with two edges identified, we perform an edge flip on the third edge of t. We underline that this assumption is not necessary, but its inclusion avoids a lot of technicalities.

The Fock-Goncharov moduli space is the the set of all positive real valued functions $\mathbb{R}_{>0}^{\triangle\cup\underline{E}} := \{\triangle \cup \underline{E} \to \mathbb{R}_{>0}\}$, and it is isomorphic to $\mathcal{T}_3^+(S)$ by Theorem 4.4.1.

For $q \in \triangle \cup \underline{E}$, the *coordinate function of* q is the map

$$\begin{aligned} q^* : \mathbb{R}_{>0}^{\triangle\cup\underline{E}} &\to \mathbb{R}_{>0}, \\ v &\mapsto v(q), \end{aligned}$$

and the q*–coordinate* is the element $\overline{q} \in \mathbb{R}_{>0}^{\triangle\cup\underline{E}}$, such that

$$\overline{q}(q') = \begin{cases} 1 & \text{if } q = q', \\ 0 & \text{otherwise.} \end{cases}$$

It is clear from the definitions that $\frac{\partial q_i^*}{\partial \overline{q}_j} = \delta_{ij}$, the Kronecker delta. For all smooth functions $f, g \in C^\infty(\mathbb{R}_{>0}^{\triangle\cup\underline{E}})$, we define

$$\{f,g\}_{\mathrm{T}} := \sum_{q_1,q_2\in\triangle\cup\underline{E}} 2\epsilon(q_1,q_2)q_1^*q_2^* \frac{\partial f}{\partial \overline{q}_1}\frac{\partial g}{\partial \overline{q}_2}, \tag{6.4.3}$$

where $\epsilon : (\triangle \cup \underline{E}) \times (\triangle \cup \underline{E}) \to \{-1, 0, 1\}$ is the skew-symmetric integral valued function defined as follows.

Given an oriented edge $e \in \underline{E}$, we will say that

- The edge e and triangle $t \in \triangle$ are *in special position* if $e \subset t$ and e, t are oriented coherently (where t has orientation induced from S).
- The edges e and $e' \in \underline{E}$ are *in special position with respect to* e if
  - There is a unique $t \in \triangle$ such that $e, e' \subset t$ and e, e' share the same starting endpoint;
  - The edge e and triangle t are in special position.

Hence we define:

$$\epsilon(e_1,e_2) = \begin{cases} 1 & \text{if } e_1, e_2 \text{ are in special position with respect to } e_1; \\ -1 & \text{if } e_1, e_2 \text{ are in special position with respect to } e_2; \\ 0 & \text{otherwise.} \end{cases}$$

$$\epsilon(e,t) = \begin{cases} 1 & \text{if } e^{-1}, t \text{ are in special position;} \\ -1 & \text{if } e, t \text{ are in special position;} \\ 0 & \text{otherwise.} \end{cases}$$

$$\epsilon(t,e) = -\epsilon(e,t), \qquad \text{and} \qquad \epsilon(t_1,t_2) = 0,$$

for all $e, e_1, e_2 \in \underline{E}$ and $t, t_1, t_2 \in \triangle$.

We remark that this is the only place where we make use of assumption (I). As mentioned before, one does not need such condition, but disposing of it makes the above definitions much more involved.

Remark 6.4.6. *The function ϵ can also be understood graphically by considering the diagram shown in Figure 6.2. Here we symbolically represent a triangle by a point in its interior and an oriented edge by a point on the edge, closer to the starting endpoint. Then $\epsilon(q_1, q_2)$ is the difference between the number of arrows from q_1 to q_2 and the number of arrows from q_2 to q_1.*

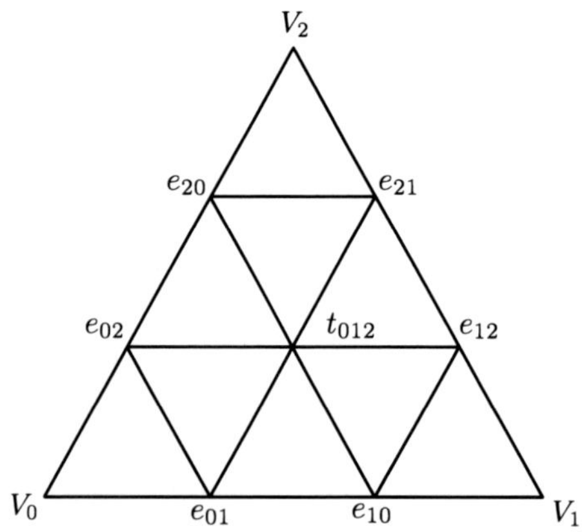

Figure 6.2: The function ϵ can be read off of a graph. We show edges connecting marked points belonging to one triangle only. Points of other triangles are connected by arrows in the same ways. Points connected by edges without arrows are not taken into account.

Proposition 6.4.7. *The bracket $\{-,-\}_{\mathrm{T}}$ is a Poisson bracket.*

Proof. Skew-symmetry, $\mathbb{R}$–bilinearity and Leibniz's product rule are simple to check. We only need to show that $\{-,-\}_{\mathrm{T}}$ satisfies the Jacobi identity, that is

$$\{f, \{g, h\}_{\mathrm{T}}\}_{\mathrm{T}} + \{g, \{h, f\}_{\mathrm{T}}\}_{\mathrm{T}} + \{h, \{f, g\}_{\mathrm{T}}\}_{\mathrm{T}} = 0,$$

for all $f, g, h \in C^{\infty}(\mathbb{R}_{>0}^{\triangle \cup \underline{E}})$. To make the computations more readable, henceforth we write $\boldsymbol{f}, \boldsymbol{g}, \boldsymbol{h}$ and simplify the notation as follows

$$\begin{aligned}\{\boldsymbol{f}, \boldsymbol{g}\}_{\mathrm{T}} := \sum_{q_1, q_2 \in \triangle \cup \underline{E}} 2\epsilon(q_1, q_2) q_1^* q_2^* \frac{\partial \boldsymbol{f}}{\partial \overline{q}_1} \frac{\partial \boldsymbol{g}}{\partial \overline{q}_2} &= \sum_{q_i, q_j \in \triangle \cup \underline{E}} \left[2\epsilon_{ij} q_i^* q_j^*\right] \frac{\partial \boldsymbol{f}}{\partial \overline{q}_i} \frac{\partial \boldsymbol{g}}{\partial \overline{q}_j} \\ &= \sum_{i,j} \omega_{ij} \partial_i \boldsymbol{f} \partial_j \boldsymbol{g}.\end{aligned}$$

Observe that $\omega_{ij} = -\omega_{ji}$. It follows that

$$\{\boldsymbol{f}, \{\boldsymbol{g}, \boldsymbol{h}\}_{\mathrm{T}}\}_{\mathrm{T}} = \sum_{i,j,k,l} \omega_{kl}\partial_k \boldsymbol{f} \left(\partial_l\omega_{ij}\partial_i \boldsymbol{g}\partial_j \boldsymbol{h} + \omega_{ij}\partial_l\partial_i \boldsymbol{g}\partial_j \boldsymbol{h} + \omega_{ij}\partial_i \boldsymbol{g}\partial_l\partial_j \boldsymbol{h}\right),$$
$$\{\boldsymbol{g}, \{\boldsymbol{h}, \boldsymbol{f}\}_{\mathrm{T}}\}_{\mathrm{T}} = \sum_{i,j,k,l} \omega_{kl}\partial_k \boldsymbol{g} \left(\partial_l\omega_{ij}\partial_i \boldsymbol{h}\partial_j \boldsymbol{f} + \omega_{ij}\partial_l\partial_i \boldsymbol{h}\partial_j \boldsymbol{f} + \omega_{ij}\partial_i \boldsymbol{h}\partial_l\partial_j \boldsymbol{f}\right),$$
$$\{\boldsymbol{h}, \{\boldsymbol{f}, \boldsymbol{g}\}_{\mathrm{T}}\}_{\mathrm{T}} = \sum_{i,j,k,l} \omega_{kl}\partial_k \boldsymbol{h} \left(\partial_l\omega_{ij}\partial_i \boldsymbol{f}\partial_j \boldsymbol{g} + \omega_{ij}\partial_l\partial_i \boldsymbol{f}\partial_j \boldsymbol{g} + \omega_{ij}\partial_i \boldsymbol{f}\partial_l\partial_j \boldsymbol{g}\right).$$

We factorise in terms of the partial derivatives of $\boldsymbol{f}, \boldsymbol{g}, \boldsymbol{h}$ to get the coefficients

$$\partial_k \boldsymbol{f}\partial_i \boldsymbol{g}\partial_j \boldsymbol{h} : \quad \sum_l \left(\omega_{kl}\partial_l\omega_{ij} + \omega_{il}\partial_l\omega_{jk} + \omega_{jl}\partial_l\omega_{ki}\right), \qquad \forall i,j,k, \tag{6.4.4}$$

$$\partial_k \boldsymbol{f}\partial_l\partial_i \boldsymbol{g}\partial_j \boldsymbol{h} : \quad \omega_{kl}\omega_{ij} + \omega_{jl}\omega_{ki}, \forall i,j,k,l, \tag{6.4.5}$$
$$\partial_k \boldsymbol{f}\partial_i \boldsymbol{g}\partial_l\partial_j \boldsymbol{h} : \quad \omega_{kl}\omega_{ij} + \omega_{il}\omega_{jk}, \forall i,j,k,l, \tag{6.4.6}$$
$$\partial_k \boldsymbol{g}\partial_i \boldsymbol{h}\partial_l\partial_j \boldsymbol{f} : \quad \omega_{kl}\omega_{ij} + \omega_{il}\omega_{jk}, \forall i,j,k,l. \tag{6.4.7}$$

When $i = l$, the coefficient in (6.4.5) is $\omega_{ki}\omega_{ij} + \omega_{ji}\omega_{ki} = 0$.
When $i \neq l$, $\partial_k \boldsymbol{f}\partial_i\partial_l \boldsymbol{g}\partial_j \boldsymbol{h} = \partial_k \boldsymbol{f}\partial_l\partial_i \boldsymbol{g}\partial_j \boldsymbol{h}$ and we can sum their coefficients to get

$$\omega_{kl}\omega_{ij} + \omega_{jl}\omega_{ki} + \omega_{ki}\omega_{lj} + \omega_{ji}\omega_{kl} = \omega_{kl}(\omega_{ij} + \omega_{ji}) + \omega_{ki}(\omega_{jl} + \omega_{lj}) = 0.$$

A similar argument applies to (6.4.6) and (6.4.7). Our claim follows from showing that the coefficients in (6.4.4) are also zero. Since

$$\frac{1}{2}\partial_l\omega_{ij} = \epsilon(q_i, q_j)\partial_l(q_i^* q_j^*) = \epsilon_{ij}\left(\delta_{il}q_j^* + \delta_{jl}q_i^*\right),$$

by substituting that and the definition of ω_{ij} in (6.4.4), we conclude that

$$\begin{aligned}&\frac{1}{2}\left(\omega_{kl}\partial_l\omega_{ij} + \omega_{il}\partial_l\omega_{jk} + \omega_{jl}\partial_l\omega_{ki}\right)\\ =&q_l^*\left[\delta_{jl}q_i^*q_k^*\left(\epsilon_{kl}\epsilon_{ij} + \epsilon_{il}\epsilon_{jk}\right) + \delta_{il}q_j^*q_k^*\left(\epsilon_{kl}\epsilon_{ij} + \epsilon_{jl}\epsilon_{ki}\right) + \delta_{kl}q_i^*q_j^*\left(\epsilon_{il}\epsilon_{jk} + \epsilon_{jl}\epsilon_{ki}\right)\right]\\ =&q_l^*\left[q_i^*q_k^*\left(\epsilon_{kj}\epsilon_{ij} + \epsilon_{ij}\epsilon_{jk}\right) + q_j^*q_k^*\left(\epsilon_{ki}\epsilon_{ij} + \epsilon_{ji}\epsilon_{ki}\right) + q_i^*q_j^*\left(\epsilon_{ik}\epsilon_{jk} + \epsilon_{jk}\epsilon_{ki}\right)\right] = 0.\end{aligned}$$

■

The next result shows that $\{-,-\}_{\mathrm{T}}$ does not depend on the triangulation T.

Theorem 6.4.8. *If* T *and* T′ *are ideal triangulations of* S*, then* $(\mathbb{R}_{>0}^{\triangle\cup\underline{E}}, \{-,-\}_{\mathrm{T}})$ *and* $(\mathbb{R}_{>0}^{\triangle'\cup\underline{E}'}, \{-,-\}_{\mathrm{T}'})$ *are naturally isomorphic Poisson manifolds.*

Proof. Let $\Phi_e : \mathbb{R}_{>0}^{\triangle\cup\underline{E}} \to \mathbb{R}_{>0}^{\triangle'\cup\underline{E}'}$ be the change of coordinates induced by flipping along an edge e of T (see §5.1.3 for details). Since Φ_e is a bijection and any two ideal triangulations of S differ by a finite sequence of edge flips, it is enough to prove that for all $f, g \in C^\infty(\mathbb{R}_{>0}^{\Delta' \cup \underline{E}'})$,

$$\{f, g\}_{\mathrm{T}'} \circ \Phi_e = \{f \circ \Phi_e, g \circ \Phi_e\}_{\mathrm{T}},$$

or equivalently, for $q_1', q_2' \in \Delta' \cup \underline{E}'$ and $q_1, q_2 \in \Delta \cup \underline{E}$,

$$\left(\sum_{q_1', q_2'} \epsilon(q_1', q_2') q_1'^* q_2'^* \frac{\partial f}{\partial \overline{q}_1'} \frac{\partial g}{\partial \overline{q}_2'} \right) \circ \Phi_e = \sum_{q_1, q_2} \epsilon(q_1, q_2) q_1^* q_2^* \frac{\partial (f \circ \Phi_e)}{\partial \overline{q}_1} \frac{\partial (g \circ \Phi_e)}{\partial \overline{q}_2}. \quad (6.4.8)$$

Denote by $(\Phi_e)_i := q_i'^* \circ \Phi_e$ the $q_i'^*$ component of Φ_e. Then

$$\mathrm{LHS} = \sum_{q_1', q_2'} \epsilon(q_1', q_2') (\Phi_e)_1 (\Phi_e)_2 \left(\frac{\partial f}{\partial \overline{q}_1'} \circ \Phi_e \right) \left(\frac{\partial g}{\partial \overline{q}_2'} \circ \Phi_e \right),$$

$$\begin{aligned} \mathrm{RHS} &= \sum_{q_1, q_2} \epsilon(q_1, q_2) q_1^* q_2^* \left[\sum_{q_1', q_2'} \left(\frac{\partial f}{\partial \overline{q}_1'} \circ \Phi_e \frac{\partial (\Phi_e)_1}{\partial \overline{q}_1} \right) \left(\frac{\partial g}{\partial \overline{q}_2'} \circ \Phi_e \frac{\partial (\Phi_e)_2}{\partial \overline{q}_2} \right) \right] \\ &= \sum_{q_1', q_2'} \sum_{q_1, q_2} \epsilon(q_1, q_2) q_1^* q_2^* \left(\frac{\partial f}{\partial \overline{q}_1'} \circ \Phi_e \frac{\partial (\Phi_e)_1}{\partial \overline{q}_1} \right) \left(\frac{\partial g}{\partial \overline{q}_2'} \circ \Phi_e \frac{\partial (\Phi_e)_2}{\partial \overline{q}_2} \right). \end{aligned}$$

Hence we rewrite equation (6.4.8) as

$$\sum_{q_1', q_2'} \left(\frac{\partial f}{\partial \overline{q}_1'} \circ \Phi_e \right) \left(\frac{\partial g}{\partial \overline{q}_2'} \circ \Phi_e \right) \cdot * = 0$$

where

$$* := \sum_{q_1, q_2} \epsilon(q_1, q_2) q_1^* q_2^* \frac{\partial (\Phi_e)_1}{\partial \overline{q}_1} \frac{\partial (\Phi_e)_2}{\partial \overline{q}_2} - \epsilon(q_1', q_2') (\Phi_e)_1 (\Phi_e)_2.$$

By differentiating the functions in section 5.1.3, one can easily check that for all i, j

$$\sum_{q_1, q_2} \epsilon(q_1, q_2) q_1^* q_2^* \frac{\partial (\Phi_e)_1}{\partial \overline{q}_1} \frac{\partial (\Phi_e)_2}{\partial \overline{q}_2} - \epsilon(q_i', q_j') (\Phi_e)_i (\Phi_e)_j = 0.$$

■

Let η be the change of coordinates induced by the sequence of edge flips between the triangulations T and T′. The following diagram commutes

$$\begin{array}{ccc} \mathcal{T}_3^+(S) & \xrightarrow{\ \phi_{\mathrm{T}}^+\ } & \mathbb{R}_{>0}^{\Delta \cup \underline{E}} \\ & \searrow^{\phi_{\mathrm{T}'}^+} & \downarrow \eta \\ & & \mathbb{R}_{>0}^{\Delta' \cup \underline{E}'} \end{array}$$

and the pull-back of $\{-,-\}_{\mathrm{T}}$ via ϕ_{T} does not depend on the coordinate system. Therefore for all $f, g \in C^\infty(\mathcal{T}_3^+(S))$ and any ideal triangulation T of S, we define

$$\{f,g\}_{FG} := \{f \circ (\phi_{\mathrm{T}}^+)^{-1}, g \circ (\phi_{\mathrm{T}}^+)^{-1}\}_{\mathrm{T}},$$

which makes $(\mathcal{T}_3^+(S), \{-,-\}_{FG})$ a Poisson manifold.

6.4.4 Compatibility of Poisson brackets

Let $\mathrm{Tr}_3 := \mathrm{Tr}(\mathfrak{X}_3(S)) \subset C^\infty(\mathfrak{X}_3(S))$ be the $\mathrm{SL}(3,\mathbb{R})$–trace algebra and let

$$\begin{aligned}\mu^*\colon C^\infty(\mathfrak{X}_3(S)) &\to C^\infty(\mathcal{T}_3^+(S)),\\ f &\mapsto f \circ \mu,\end{aligned}$$

be the injective map induced by the monodromy operator $\mu\colon \mathcal{T}_3^+(S) \to \mathfrak{X}_3(S)$. This section is devoted to the proof of the following result.

Theorem 6.4.9. *The restriction of* μ^* *to* Tr_3 *is a Poisson homomorphism, i.e. for all* $f, g \in \mathrm{Tr}_3$,

$$2\ \mu^* \circ \{f,g\}_{Gol} = \{\mu^* \circ f, \mu^* \circ g\}_{FG}. \tag{6.4.9}$$

The strategy we adopt consists of the following 5 steps. First we apply Theorem 4.4.1 to reword the problem in terms of $\{-,-\}_{\mathrm{T}}$ for a fixed triangulation T of S. That allows us to compare the definition of $\{-,-\}_{\mathrm{T}}$ in (6.4.3) with the formulation of $\{-,-\}_{Gol}$ in (6.4.1) and just focus on representatives α, β of immersed oriented closed curves on S in general position. Next, we break up α and β into pieces and focus our attention on *tunnels and galleries*, where curves travel next to each other (see §6.4.6 for details). We then show in §6.4.7 how the global quantity $\{\mathrm{tr}_\alpha \circ \mu, \mathrm{tr}_\beta \circ \mu\}_{\mathrm{T}}$ translates into a local behaviour $\{\mathrm{tr}_\alpha, \mathrm{tr}_\beta\}_{Gol}$ at the intersection points of α and β. A special case when α and β are *parallel* is treated in §6.4.8. All of the pieces are put together to conclude the proof in §6.4.9.

6.4.5 Proof part 1: set up

Fix an orientation, an ideal triangulation T of S and the corresponding isomorphism $\phi_{\mathrm{T}}^+ : \mathcal{T}_3^+(S) \to \mathbb{R}_{>0}^{\Delta\cup E}$. Let $(\mu \circ (\phi_{\mathrm{T}}^+)^{-1})^*\colon C^\infty(\mathfrak{X}_3(S)) \to C^\infty(\mathbb{R}_{>0}^{\Delta\cup E})$ be the map $(\mu\circ(\phi_{\mathrm{T}}^+)^{-1})^*(f) = f\circ\mu\circ(\phi_{\mathrm{T}}^+)^{-1}$, then μ^* is a Poisson homomorphism on Tr_3 if and only if $(\mu \circ (\phi_{\mathrm{T}}^+)^{-1})^*$ is a Poisson homomorphism on Tr_3. Henceforth we identify $\mathcal{T}_3^+(S)$ with $\mathbb{R}_{>0}^{\Delta\cup E}$ and denote by μ, μ^* the maps $\mu \circ (\phi_{\mathrm{T}}^+)^{-1}, (\mu \circ (\phi_{\mathrm{T}}^+)^{-1})^*$.

Recall that Tr_3 is the sub-algebra of $C^\infty(\mathfrak{X}_3(S))$ generated by elements of the form tr_α for $\alpha \in \pi_1(S)$. By Theorem 6.4.5 and the defining properties of Poisson brackets, it is enough to show that for all representatives $\alpha, \beta \subset S$ of immersed oriented closed curves in general position,

$$\begin{aligned}
&\sum_{p\in\alpha\cap\beta} \epsilon(p,\alpha,\beta)\left(\operatorname{tr}_{\alpha_p\beta_p}\circ\mu - \frac{1}{3}(\operatorname{tr}_\alpha\circ\mu)(\operatorname{tr}_\beta\circ\mu)\right)\\
&= \sum_{q_1,q_2\in\triangle\cup\underline{E}} \epsilon(q_1,q_2)q_1^* q_2^* \frac{\partial(\operatorname{tr}_\alpha\circ\mu)}{\partial\overline{q}_1}\frac{\partial(\operatorname{tr}_\beta\circ\mu)}{\partial\overline{q}_2}.
\end{aligned} \tag{6.4.10}$$

Here α and β are assumed to be in general position both with respect to each other and with respect to T.

We start by recalling some notation from §5.2. For $\gamma\in\pi_1(S)$, we showed in Lemma 5.2.1 that there exists a finite sequence $(e_0^\gamma\cdot\dots\cdot e_n^\gamma)$ of $n+1$ oriented edges $e_i^\gamma\in\underline{E}(\mathcal{G})$ such that $e_0^\gamma\cdot e_1^\gamma\cdot\dots\cdot e_n^\gamma\subset\mathcal{G}$ is the development of a loop in $\mathcal{G}_0$ homotopic equivalent to γ. Let $(s_0^\gamma\cdot\dots\cdot s_n^\gamma)$ be such a loop, where $s_i^\gamma\in\underline{E}(\mathcal{G}_0)$ corresponds to $e_i^\gamma\in\underline{E}(\mathcal{G})$. Then we showed in Theorem 5.2.4 that $\mu(x)(\gamma)$ is the conjugacy class of a product $S_0^\gamma\cdot S_1^\gamma\cdots S_n^\gamma$, where S_i^γ's are matrices of the type T,T^{-1} or E. More precisely, the matrix S_i^γ is of type T,T^{-1} (respectively type E) if and only if s_i^γ is a $\triangle$–edge (resp. $\underline{E}$–edge), as described at the beginning of §5.2. Furthermore, we can assume that s_i^γ and s_{i+1}^γ are of different type for all $i\in\mathbb{Z}_{n+1}$.

For the remainder of this section we will work with fixed sequences e_i^α, e_i^β, loops s_i^α, s_i^β and matrices $S_i^\alpha, S_i^\beta\in\mathrm{SL}(3,\mathbb{R})$ for α and β respectively.

For convenience, we provide here a simple result which will be used later on.

Lemma 6.4.10. *Let $T(z), E(x,y)\in\mathrm{SL}(3,\mathbb{R})$ be the parametric matrices defined in Lemma 5.2.3, and let*

$$M_1 := \begin{pmatrix} -2 & 0 & 0\\ 0 & 1 & 0\\ 0 & 0 & 1\end{pmatrix},\qquad M_2 := \begin{pmatrix} -1 & 0 & 0\\ 0 & -1 & 0\\ 0 & 0 & 2\end{pmatrix},$$

$$M_3 := \begin{pmatrix} 2 & 3 & 0\\ 0 & -1 & 0\\ 0 & 0 & -1\end{pmatrix},\qquad M_4 := \begin{pmatrix} 1 & 0 & 0\\ 0 & 1 & 0\\ 0 & -3 & -2\end{pmatrix}.$$

Then

$$\frac{\partial E(x,y)}{\partial x} = \frac{1}{3x}E(x,y)\cdot M_1,\qquad \frac{\partial E(x,y)}{\partial y} = \frac{1}{3y}E(x,y)\cdot M_2,$$

$$\frac{\partial T(z)}{\partial z} = \frac{1}{3z}T(z)\cdot M_3,\qquad \frac{\partial (T(z)^{-1})}{\partial z} = \frac{1}{3z}T(z)^{-1}\cdot M_4.$$

6.4.6 Proof part 2: tunnels and galleries

The function $\{f,g\}_{FG}$ does not depend on the chosen triangulation T so we may assume, without loss of generality, that:

(I) *Every edge in* T *is contained in two distinct triangles, or equivalently, every triangle has three distinct edges.*

As before, we underline that this assumption is not necessary and we will specify where we use it.

Denote by $\partial_i f := \frac{\partial f}{\partial q_i}$, then by the product rule

$$\partial_i(\operatorname{tr}_\gamma \circ \mu) = \partial_i(\operatorname{tr} \mu(\gamma)) = \operatorname{tr}(\partial_i \mu(\gamma)) = \sum_j \operatorname{tr}\left(S_0^\gamma \cdots \partial_i S_j^\gamma \cdots S_n^\gamma\right).$$

Denote $s_{j,k}^\gamma := S_0^\gamma \cdots \partial_j S_k^\gamma \cdots S_n^\gamma$ for all j, k. We remind the reader that s_k^γ is a segment of a loop homotopic to γ, so the notation is chosen to emphasize that the partial derivative is "taken at s_k^γ".

It follows that

$$\partial_1(\operatorname{tr}_\alpha \circ \mu) \cdot \partial_2(\operatorname{tr}_\beta \circ \mu) = \sum_{i,j} \operatorname{tr}(s_{1,i}^\alpha) \cdot \operatorname{tr}(s_{2,j}^\beta),$$

and the right hand side of (6.4.10) can be reformulated as

$$\sum_{i,j} \sum_{q_1, q_2 \in \Delta \cup \underline{E}} \epsilon(q_1, q_2) q_1^* q_2^* \left(\operatorname{tr}(s_{1,i}^\alpha) \cdot \operatorname{tr}(s_{2,j}^\beta)\right). \tag{6.4.11}$$

Remark 6.4.11. *Lemma 6.4.10 implies that*

$$s_{j,k}^\gamma = \frac{1}{3q_j^*}\left(S_0^\gamma \cdots S_k^\gamma \cdot X \cdots S_n^\gamma\right),$$

where X is:

i) M_1 or M_2, if s_k^γ is an $\underline{E}$–edge crossing an oriented edge of T *with edge ratio q_j;*
ii) M_3 or M_4, if s_k^γ is a Δ–edge contained in a triangle of T *with triple ratio q_j;*
iii) the zero matrix otherwise.

We are now going to investigate the set of pairs (s_i^α, s_j^β) and their contribution

$$C(s_i^\alpha, s_j^\beta) := \sum_{q_1, q_2 \in \Delta \cup \underline{E}} \epsilon(q_1, q_2) q_1^* q_2^* \left(\operatorname{tr}(s_{1,i}^\alpha) \cdot \operatorname{tr}(s_{2,j}^\beta)\right)$$

in the above sum (6.4.11).

Clearly $C(s_i^\alpha, s_j^\beta) = 0$ when s_i^α and s_j^β do not intersect a common triangle. If α and β never cross a common triangle, then they must be disjoint, hence both right hand side and left hand side in (6.4.10) are 0.

For that reason, we assume from now on that

(II) *There exist segments s_i^α and s_j^β intersecting a common triangle t of* T.

When t is a common triangle to two segments s_i^α and s_j^β it follows from assumption (I) that there is a unique (oriented) connected component $t(s_i^\alpha)$ (resp. $t(s_j^\beta)$) of

$t \cap \alpha$ (resp. $t \cap \beta$) intersecting s_i^α (resp. s_j^β). We will say that t has the *configuration type* of:

- a *starting triangle* (resp. *ending* triangle) with respect to s_i^α and s_j^β, if $t(s_i^\alpha)$ and $t(s_j^\beta)$ cross a unique common edge of t and $t(s_i^\alpha)$ leaves (resp. enters) t from such edge;
- a *transition triangle* with respect to s_i^α and s_j^β, if $t(s_i^\alpha)$ and $t(s_j^\beta)$ cross the same two edges of t.

Up to interchanging α and β, there are only 7 possible configurations. They are listed in Figure 6.3. We denote by jS the configuration of type j where α and β have been interchanged. This produces 5 new configurations because configurations 5 and 6 are the same as configurations $5S$ and $6S$.

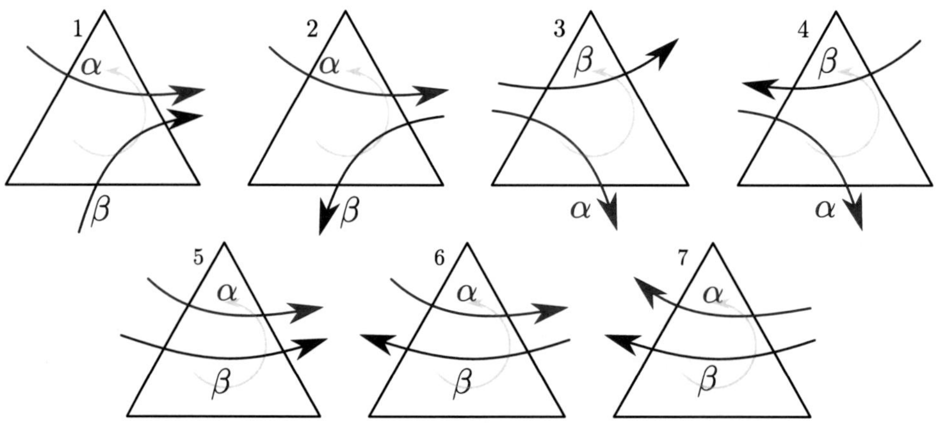

Figure 6.3: Triangle configuration types.

Furthermore, we will say that $w_\gamma = (w_0^\gamma \cdots w_k^\gamma)$ is a *walk along* $(s_0^\gamma \cdots s_n^\gamma)$ of length $k+1 \in \mathbb{N}$ based at s_j^γ if either

$$w_i^\gamma = s^\gamma_{[j+i]_{n+1}}, \qquad \text{or} \qquad w_i^\gamma = s^\gamma_{[j-i]_{n+1}},$$

where $[m]_{n+1} := m \mod (n+1)$. When γ is represented by a loop $(s_0^\gamma \cdots s_n^\gamma)$, we will abbreviate the notation by saying that w_γ is a *walk along* γ.

Remark 6.4.12. *Broadly speaking, w_γ is a walk along $(s_0^\gamma \cdots s_n^\gamma)$ starting at s_j^γ which might wrap around the curve multiple times, possibly backwards with respect to its orientation. In particular, $(s_0^\gamma \cdot s_1^\gamma \cdots s_n^\gamma)$ and $(s_0^\gamma \cdot s_n^\gamma \cdots s_1^\gamma)$ are walks of length $n+1$ based at s_0^γ in different directions.*

For the moment, we will make the extra assumption that

($\star$) The curves α *and* β *are non-parallel.*

That is $[\alpha] \neq [\beta^m]$ for all $m \in \mathbb{Z}$. We come back to this particular case later in §6.4.8.

Since α and β are closed non-parallel curves, if t is a common triangle to two segments s_i^α and s_j^β then there exist:

- a sequence of adjacent (not necessarily distinct) triangles $(t_0, \ldots, t_q) \subset \mathrm{T}$ containing t;
- a walk $w_\alpha := (w_0^\alpha \cdot \cdots \cdot w_{2q+2}^\alpha)$ along α containing s_i^α;
- a walk $w_\beta := (w_0^\beta \cdot \cdots \cdot w_{2q+2}^\beta)$ along β containing s_j^β;

such that t_0 is a starting triangle with respect to the pairs $(w_i^\alpha, w_i^\beta)_{0 \leqslant i \leqslant 2}$, t_q is an ending triangle with respect to the pairs $(w_i^\alpha, w_i^\beta)_{2q \leqslant i \leqslant 2q+2}$, and t_j is a transition triangle with respect to the pairs $(w_i^\alpha, w_i^\beta)_{2j \leqslant i \leqslant 2j+2}$ for all $1 \leqslant j \leqslant q-1$.

In this case we will say that (w_α, w_β) is a *tunnel through* $(t_0, \ldots, t_q)$ *containing* (s_i^α, s_j^β). Furthermore, for all $k \in \{0 \ldots q\}$ there is a preferred triple of pairs $(w_i^\alpha, w_i^\beta)_{2k \leqslant i \leqslant 2k+2}$ and a triangle t_k associated to it, which will be called the k–*th gallery* of the tunnel. By definition every j–th and $(j+1)$–th gallery share a common pair $(w_{2j+2}^\alpha, w_{2j+2}^\beta)$.

Remark 6.4.13. *A good way to visualise a tunnel is to pass to a universal cover of the surface. If (w_α, w_β) is a tunnel through $(t_0, \ldots, t_q)$ and we lift α and β to the universal cover $\widetilde{S}$ based at a some point p in any of the triangles t_i, then the two curves in $\widetilde{S}$ travel through common triangles only along the tunnel while apart from each other everywhere else.*

Henceforth, when we talk about a tunnel containing the pair (s_i^α, s_j^β), we will further assume that s_i^α and s_j^β are contained in the same gallery. In this way, for every pair intersecting a common triangle there is a unique tunnel containing them. We stress again that the assumption $(\star)$ is necessary for the existence of at least one tunnel, as the reader can be easily convinced by taking simple closed curves $\alpha = \beta$.

6.4.7 Proof part 3: encoding the contribution along a tunnel

Let (w_α, w_β) be a tunnel τ through $(t_0, \ldots, t_q)$. We now develop a handy way to encode the *contribution* $C_k(\tau)$ *of the* k–*th gallery of* τ, that is the quantity

$$C_k(\tau) := \sum_{2k \leqslant i,j \leqslant 2k+2} \left(\sum_{q_1, q_2 \subset t_k} \epsilon(q_1, q_2) q_1^* q_2^* \left(\operatorname{tr}(w_{1,i}^\alpha) \cdot \operatorname{tr}(w_{2,j}^\beta) \right) \right). \tag{6.4.12}$$

Following the notation of the previous paragraph, let W_i^γ be the matrix in the product $S_0^\gamma \cdot S_1^\gamma \cdots S_n^\gamma$ corresponding to the segment w_i^γ in $(s_0^\gamma \cdot \cdots \cdot s_n^\gamma)$. Then for all $k \in \{0, \dots, q\}$, $m \in \{0, 1, 2\}$ and $\gamma \in \{\alpha, \beta\}$,

$$\begin{aligned} q_i^* \operatorname{tr}(w^{\gamma}_{i,2k+m}) &= q_i^* \operatorname{tr}(S_0^{\gamma} \cdots \partial_i W^{\gamma}_{2k+m} \cdots S_n^{\gamma}) \\ &= \frac{1}{3} \operatorname{tr}(S_0^{\gamma} \cdots W^{\gamma}_{2k+m} \cdot \widetilde{X} \cdots S_n^{\gamma}) \qquad \text{see Remark 6.4.11,} \\ &= \frac{1}{3} \operatorname{tr}(W^{\gamma}_{2k+m} \cdot \widetilde{X} \cdot \cdots \cdot S_n^{\gamma} \cdot S_0^{\gamma} \cdots) \text{ by conjugation invariance of tr,} \\ &= \frac{1}{3} \operatorname{tr}(X \cdot W^{\gamma}_{2k+1} \cdot \cdots \cdot S_n^{\gamma} \cdot S_0^{\gamma} \cdots), \end{aligned}$$

where $\widetilde{X} = \widetilde{X}(\gamma, q_i, k, m)$ is a matrix dependent on (γ, q_i, k, m) and

- for $\gamma = \alpha$,

$$X = \left(\prod_{i=1}^{m} W^{\alpha}_{2k+i} \right) \cdot \widetilde{X} \cdot \left(\prod_{i=1}^{m} W^{\alpha}_{2k+i} \right)^{-1};$$

- for $\gamma = \beta$, if β is oriented as α along the tunnel,

$$X = \left(\prod_{i=1}^{m} W^{\beta}_{2k+i} \right) \cdot \widetilde{X} \cdot \left(\prod_{i=1}^{m} W^{\beta}_{2k+i} \right)^{-1};$$

  otherwise

$$X = \left(\prod_{i=0}^{1-m} W^{\beta}_{2k+1-i} \right) \cdot \widetilde{X} \cdot \left(\prod_{i=0}^{1-m} W^{\beta}_{2k+1-i} \right)^{-1}.$$

  In short we can write

$$q_i^* \operatorname{tr}(w^{\gamma}_{i,2k+m}) = \frac{1}{3} \operatorname{tr}(X \cdot \mathbf{S}(w^{\gamma}_{2k+1})),$$

where $X = X(\gamma, q_i, k, m)$ is a matrix dependent on (γ, q_i, k, m) and

$$\mathbf{S}(w^{\gamma}_{2k+1}) := W^{\gamma}_{2k+1} \cdot \cdots \cdot S_n^{\gamma} \cdot S_0^{\gamma} \cdots,$$

is the matrix conjugate to $S_0^{\gamma} \cdots W^{\gamma}_{2k+1} \cdots S_n^{\gamma}$ obtained by moving all the matrices on the left side of W^{γ}_{2k+1} to the right side of S_n^{γ}.

Lemma 6.4.14. *Let $X, Y \in M_{3\times3}(\mathbb{R})$ be real matrices and let $\mathbf{E}_{ij} \in M_{3\times3}(\mathbb{R})$ be the matrix with (i, j)–th entry 1 and zeros everywhere else. Then*

$$\operatorname{tr}(X \cdot Y) = \sum_{i,j} x_{ij} \operatorname{tr}(\mathbf{E}_{ij} \cdot Y),$$

where x_{ij} is the (i, j)–th entry of X.

By Lemma 6.4.14 and the above discussion we can conclude that

$$
\begin{aligned}
9C_k(\tau) &= 9 \sum_{0\leqslant m_1,m_2\leqslant 2}\left(\sum_{q_1,q_2\subset t_k} \epsilon(q_1,q_2)q_1^*q_2^*\left(\mathrm{tr}(w^{\alpha}_{1,2k+m_1})\cdot \mathrm{tr}(w^{\beta}_{2,2k+m_2})\right)\right)\\
&= \sum_{0\leqslant m_1,m_2\leqslant 2}\left(\sum_{q_1,q_2\subset t_k} \epsilon(q_1,q_2)\left(\mathrm{tr}\left(X(\alpha,q_1,k,m_1)\cdot \mathbf{S}(w^{\alpha}_{2k+1})\right)\cdot *_1\right)\right)\\
&= \sum_{l,m}\sum_{0\leqslant m_1,m_2\leqslant 2}\left(\sum_{q_1,q_2\subset t_k} \epsilon(q_1,q_2)\left(x^{(\alpha,q_1,k,m_1)}_{lm}\,\mathrm{tr}\left(\mathbf{E}_{lm}\cdot \mathbf{S}(w^{\alpha}_{2k+1})\right)\cdot *_1\right)\right)\\
&= \sum_{l,m} *_2\cdot \sum_{0\leqslant m_1,m_2\leqslant 2}\left(\sum_{q_1,q_2\subset t_k} \epsilon(q_1,q_2)\left(x^{(\alpha,q_1,k,m_1)}_{lm}\cdot *_1\right)\right)\\
&= \sum_{l,m} *_2\cdot \mathrm{tr}\left(\left(\sum_{0\leqslant m_1,m_2\leqslant 2}\sum_{q_1,q_2\subset t_k} \epsilon(q_1,q_2)x^{(\alpha,q_1,k,m_1)}_{lm}X(\beta,q_2,k,m_2)\right)\cdot \mathbf{S}(w^{\beta}_{2k+1})\right)\\
&= \sum_{l,m} *_2\cdot \mathrm{tr}(U^{(l,m)}\cdot \mathbf{S}(w^{\beta}_{2k+1})).
\end{aligned}
$$

where

$$
*_1 := \mathrm{tr}\left(X(\beta,q_2,k,m_2)\cdot \mathbf{S}(w^{\beta}_{2k+1})\right); \tag{6.4.13}
$$

$$
*_2 := \mathrm{tr}\left(\mathbf{E}_{lm}\cdot \mathbf{S}(w^{\alpha}_{2k+1})\right); \tag{6.4.14}
$$

$$
U^{(l,m)} := \sum_{0\leqslant m_1,m_2\leqslant 2}\sum_{q_1,q_2\subset t_k} \epsilon(q_1,q_2)x^{(\alpha,q_1,k,m_1)}_{lm}X(\beta,q_2,k,m_2). \tag{6.4.15}
$$

and $x^{(\alpha,q_1,k,m_1)}_{lm}$ is the (l,m)–th entry of $X(\alpha,q_1,k,m_1)$.

Let $\mathbf{M}_k(\tau)\in M_{3\times 3}(M_{3\times 3}(\mathbb{R}))$ be the 3×3 matrix whose (l,m)–th entry is the 3×3 matrix $U^{(l,m)}$. Then $\mathbf{M}_k(\tau)$ encodes $C_k(\tau)$ in the sense that

$$
C_k(\tau) = \frac{1}{9}\sum_{l,m}\mathrm{tr}(\mathbf{E}_{lm}\cdot \mathbf{S}(w^{\alpha}_{2k+1}))\cdot \mathrm{tr}((\mathbf{M}_k(\tau))_{lm}\cdot \mathbf{S}(w^{\beta}_{2k+1})).
$$

The matrix $\mathbf{M}_k(\tau)$ is unique and uniquely determines $C_k(\tau)$, thus we call $\mathbf{M}_k(\tau)$ the *contribution matrix* of the k–th gallery of τ.

Example 1. *We show how to compute the contribution matrix for a gallery with an explicit example. Consider the* 0*–th gallery of a tunnel* τ *where* t_0 *is a starting triangle of type* 1 *with respect to the pairs* $(w^{\alpha}_i, w^{\beta}_i)_{0\leqslant i\leqslant 2}$*, as in Figure 6.4.*

For some matrices A, B*,*

$$
\begin{aligned}
\mu(x)(\alpha) &= [E(a,b)\cdot T(z)^{-1}\cdot E(c,d)\cdot A],\\
\mu(x)(\beta) &= [E(e,f)\cdot T(z)\cdot E(c,d)\cdot B].
\end{aligned}
$$

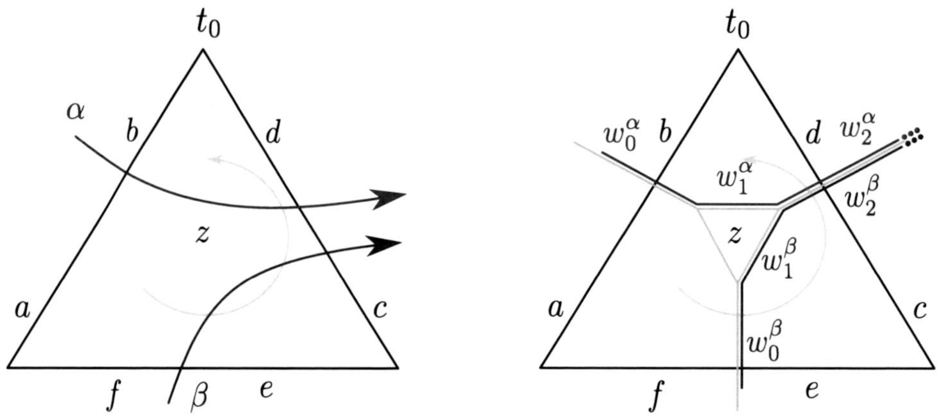

Figure 6.4: Example of contribution along a gallery with triangle of type 1.

We apply Lemma 6.4.10 and the conjugacy invariance of the trace operator to obtain:

$$q_j^* \operatorname{tr}(w_{j,k}^\alpha) = \frac{1}{3}\operatorname{tr}(X^{(\alpha,q_j,k)} \cdot T(z)^{-1} \cdot E(c,d) \cdot A \cdot E(a,b)),$$
$$q_j^* \operatorname{tr}(w_{j,k}^\beta) = \frac{1}{3}\operatorname{tr}(X^{(\beta,q_j,k)} \cdot T(z) \cdot E(c,d) \cdot B \cdot E(e,f)),$$

where explicitly (see Lemma 6.4.10 for the notation):

$$X^{(\alpha,a,0)} = M_1 = X^{(\beta,e,0)}, \qquad X^{(\alpha,c,2)} = T(z)^{-1} \cdot E(c,d) \cdot M_1 \cdot E(c,d)^{-1} \cdot T(z),$$
$$X^{(\alpha,b,0)} = M_2 = X^{(\beta,f,0)}, \qquad X^{(\beta,c,2)} = T(z) \cdot E(c,d) \cdot M_1 \cdot E(c,d)^{-1} \cdot T(z),$$
$$X^{(\alpha,z,1)} = T(z)^{-1} \cdot M_4 \cdot T(z), \quad X^{(\alpha,d,2)} = T(z)^{-1} \cdot E(c,d) \cdot M_2 \cdot E(c,d)^{-1} \cdot T(z),$$
$$X^{(\beta,z,1)} = T(z) \cdot M_3 \cdot T(z)^{-1}, \quad X^{(\beta,d,2)} = T(z) \cdot E(c,d) \cdot M_2 \cdot E(c,d)^{-1} \cdot T(z),$$

$X^{(\alpha,q_j,k)} = X^{(\beta,q_j,k)} =$ *the zero matrix otherwise.*

We can now compute the entries of $\mathbf{M}_0(\tau)$ *from equation (6.4.15), for example:*

$$\begin{aligned}
U^{(1,1)} =& x_{11}^{(\alpha,a,0)} \left(\epsilon(a,f)X^{(\beta,f,0)} + \epsilon(a,z)X^{(\beta,z,1)}\right) \\
&+ x_{11}^{(\alpha,b,0)} \left(\epsilon(b,d)X^{(\beta,d,2)} + \epsilon(b,z)X^{(\beta,z,1)}\right) \\
&+ x_{11}^{(\alpha,z,1)} \left(\epsilon(z,f)X^{(\beta,f,0)} + \epsilon(z,e)X^{(\beta,e,0)} + \epsilon(z,c)X^{(\beta,c,2)} + \epsilon(z,d)X^{(\beta,d,2)}\right) \\
&+ x_{11}^{(\alpha,c,2)} \left(\epsilon(c,e)X^{(\beta,e,0)} + \epsilon(c,z)X^{(\beta,z,1)}\right) \\
&+ x_{11}^{(\alpha,d,2)} \left(\epsilon(d,z)X^{(\beta,z,1)}\right).
\end{aligned}$$

It follows that the contribution matrix of the 0*–th gallery is*

$$\mathbf{M}_0(\tau) = \begin{pmatrix} \mathbf{M}_0^{1,1} & \mathbf{M}_0^{1,2} & \mathbf{M}_0^{1,3} \\ \mathbf{0} & \mathbf{M}_0^{2,2} & \mathbf{M}_0^{2,3} \\ \mathbf{0} & \mathbf{0} & \mathbf{M}_0^{3,3} \end{pmatrix},$$

where

$$\mathbf{M}_0^{1,1} = \begin{pmatrix} 3 & 0 & 0 \\ -6 & -3 & 0 \\ 3(z+2) & 3(z+1) & 0 \end{pmatrix}, \qquad \mathbf{M}_0^{2,2} = \begin{pmatrix} 0 & 0 & 0 \\ 3 & 3 & 0 \\ -3(2z+1) & -6(z+1) & -3 \end{pmatrix},$$

$$\mathbf{M}_0^{1,2} = \begin{pmatrix} \frac{3(z+1)}{z} & 0 & 0 \\ -9 & -\frac{6(z+1)}{z} & 0 \\ 9(z+1) & \frac{9(z+1)^2}{z} & \frac{3(z+1)}{z} \end{pmatrix}, \quad \mathbf{M}_0^{2,3} = \begin{pmatrix} 3 & 0 & 0 \\ 0 & 3 & 0 \\ 0 & -9 & -6 \end{pmatrix},$$

$$\mathbf{M}_0^{1,3} = \begin{pmatrix} \frac{3(z-1)}{z} & 0 & 0 \\ 0 & -\frac{3(z+2)}{z} & 0 \\ 0 & \frac{9(z+1)}{z} & \frac{3(2z+1)}{z} \end{pmatrix}, \quad \mathbf{M}_0^{3,3} = \begin{pmatrix} -3 & 0 & 0 \\ 3 & 0 & 0 \\ 3(z-1) & 3(z+1) & 3 \end{pmatrix}.$$

Therefore the contribution of the 0*–th gallery is:* $C_0(\tau) = \frac{1}{9}\sum_{l,m} \operatorname{tr}(*_1) \cdot \operatorname{tr}(*_2)$, *where*

$$*_1 := \mathbf{E}_{lm} \cdot T(z)^{-1} \cdot E(c,d) \cdot A \cdot E(a,b),$$
$$*_2 := (\mathbf{M}_0(\tau))_{lm} \cdot T(z) \cdot E(c,d) \cdot B \cdot E(e,f).$$

Lemma 6.4.15. *Let* $C_{k-1}(\tau), C_k(\tau)$ *and* $\mathbf{M}_{k-1}(\tau), \mathbf{M}_k(\tau)$ *be the contributions and contribution matrices of two consecutive galleries in a tunnel* τ. *Let*

- $a_{ij}^{(l,m)}$ *be the* (i,j)*–th entry of the matrix*

$$A^{(l,m)} := (W_{2k-1}^{\alpha} \cdot W_{2k}^{\alpha})^{-1} \cdot \mathbf{E}_{lm} \cdot W_{2k-1}^{\alpha} \cdot W_{2k}^{\alpha};$$

- *if* α, β *are oriented in the same direction along the galleries,*

$$B^{(l,m)} := (W_{2k-1}^{\beta} \cdot W_{2k}^{\beta})^{-1} \cdot (\mathbf{M}_{k-1}(\tau))_{lm} \cdot W_{2k-1}^{\beta} \cdot W_{2k}^{\beta},$$

otherwise

$$B^{(l,m)} := W_{2k+1}^{\beta} \cdot W_{2k}^{\beta} \cdot (\mathbf{M}_{k-1}(\tau))_{lm} \cdot (W_{2k+1}^{\beta} \cdot W_{2k}^{\beta})^{-1}.$$

Then

$$C_{k-1}(\tau) + C_k(\tau) = \frac{1}{9}\sum_{l,m} \operatorname{tr}(\mathbf{E}_{lm} \cdot \mathbf{S}(w_{2k+1}^{\alpha})) \cdot \operatorname{tr}((\mathbf{M}_{(k-1,k)}(\tau))_{lm} \cdot \mathbf{S}(w_{2k+1}^{\beta})),$$

where $\mathbf{M}_{(k-1,k)}(\tau) \in M_{3\times 3}(M_{3\times 3}(\mathbb{R}))$ *is the matrix with* (l,m)*–th entry*

$$(\mathbf{M}_{(k-1,k)}(\tau))_{lm} = \sum_{i,j} a_{lm}^{(i,j)} B^{(i,j)} + (\mathbf{M}_k(\tau))_{lm}.$$

Proof.

$$\begin{aligned}
C_{k-1}(\tau) &= \frac{1}{9}\sum_{l,m} \operatorname{tr}(\mathbf{E}_{lm}\cdot \mathbf{S}(w^{\alpha}_{2k-1}))\cdot \operatorname{tr}((\mathbf{M}_{k-1}(\tau))_{lm}\cdot \mathbf{S}(w^{\beta}_{2k-1}))\\
&= \frac{1}{9}\sum_{l,m} \operatorname{tr}(A^{(l,m)}\cdot \mathbf{S}(w^{\alpha}_{2k+1}))\cdot \operatorname{tr}(B^{(l,m)}\cdot \mathbf{S}(w^{\beta}_{2k+1}))\\
&= \frac{1}{9}\sum_{l,m}\sum_{i,j} a^{(l,m)}_{ij} \operatorname{tr}(\mathbf{E}_{ij}\cdot \mathbf{S}(w^{\alpha}_{2k+1}))\cdot \operatorname{tr}(B^{(l,m)}\cdot \mathbf{S}(w^{\beta}_{2k+1}))\\
&= \frac{1}{9}\sum_{i,j} \operatorname{tr}(\mathbf{E}_{ij}\cdot \mathbf{S}(w^{\alpha}_{2k+1}))\cdot \operatorname{tr}(\sum_{l,m} a^{(l,m)}_{ij}\cdot B^{(l,m)}\cdot \mathbf{S}(w^{\beta}_{2k+1}))\\
&= \frac{1}{9}\sum_{l,m} \operatorname{tr}(\mathbf{E}_{lm}\cdot \mathbf{S}(w^{\alpha}_{2k+1}))\cdot \operatorname{tr}(\sum_{i,j} a^{(i,j)}_{lm}\cdot B^{(i,j)}\cdot \mathbf{S}(w^{\beta}_{2k+1})).
\end{aligned}$$

We remark that the second equality is obtained by conjugation, the third one follows from Lemma 6.4.14 and the last one is obtained by renaming the indices. Since by definition

$$C_k(\tau) = \frac{1}{9}\sum_{l,m} \operatorname{tr}(\mathbf{E}_{lm}\cdot \mathbf{S}(w^{\alpha}_{2k+1}))\cdot \operatorname{tr}((\mathbf{M}_k(\tau))_{lm}\cdot \mathbf{S}(w^{\beta}_{2k+1})),$$

then the result follows. ■

Remark 6.4.16. *From a computational point of view, it is interesting to observe that the matrix* $\sum_{i,j} a^{(i,j)}_{lm}\cdot B^{(i,j)}$ *is the* (l,m)*–th entry of* $(W^{\alpha}_{2k-1}\cdot W^{\alpha}_{2k})^{-1}\cdot \mathbf{B}\cdot W^{\alpha}_{2k-1}\cdot W^{\alpha}_{2k}$*, where* $\mathbf{B}\in M_{3\times 3}(M_{3\times 3}(\mathbb{R}))$ *is the matrix with matrix entries* $B^{(i,j)}$.

For each starting triangle configuration, there are only two compatible ending triangles, which makes a total of 8 possible tunnels without transition triangles (cf. Figure 6.5).

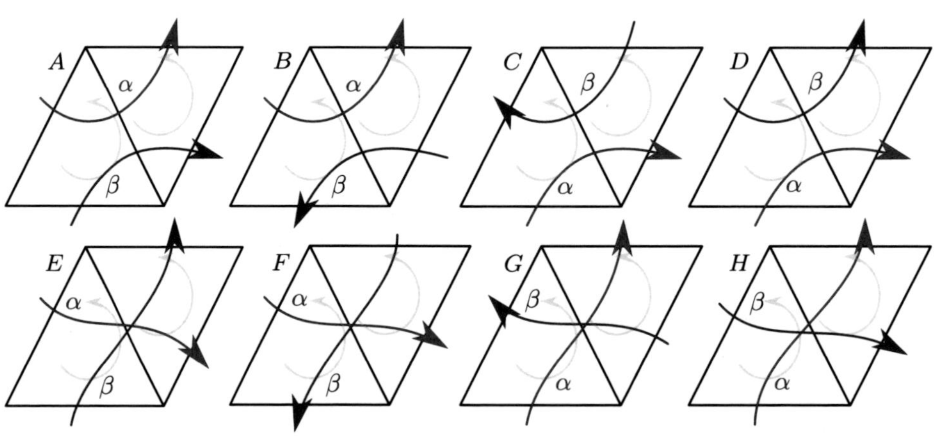

Figure 6.5: Length two tunnels.

Lemma 6.4.17. *The contribution along tunnels of length two equals*

$$\epsilon(p,\alpha,\beta)\left(\mathrm{tr}_{\alpha_p\beta_p}\circ\mu-\frac{1}{3}(\mathrm{tr}_\alpha\circ\mu)(\mathrm{tr}_\beta\circ\mu)\right),$$

where p is the point of intersection (if any) of α and β in the tunnel.

Proof. The same technique applies to each case, hence we pay our attention to just one of them. We will briefly explain at the end how to treat the others.

Let $(w_i^\alpha, w_i^\beta)_{0\leqslant i\leqslant 4}$ be a tunnel τ through (t_0,t_1) of type C (as in Figure 6.5, namely t_0 is a starting triangle of type $4S$ (which is type 4 with α and β switched) with respect to the pairs $(w_i^\alpha, w_i^\beta)_{0\leqslant i\leqslant 2}$, and t_1 is an ending triangle of type 4 for $(w_i^\alpha, w_i^\beta)_{2\leqslant i\leqslant 4}$. According to the labels shown in Figure 6.6, we have

$$\begin{aligned}\mu(x)(\alpha)&=[E(f,e)\cdot T(z)\cdot E(d,c)\cdot T(w)\cdot E(g,h)\cdot A],\\ \mu(x)(\beta)&=[E(j,i)\cdot T(w)\cdot E(c,d)\cdot T(z)\cdot E(a,b)\cdot B],\end{aligned}$$

for some matrices A, B.

The contribution matrices $\mathbf{M}_0(\tau),\mathbf{M}_1(\tau)$ at t_0,t_1 are:

$$\mathbf{M}_0(\tau)=\begin{pmatrix}\mathbf{M}_0^{1,1}&\mathbf{M}_0^{1,2}&\mathbf{M}_0^{1,3}\\ \mathbf{0}&\mathbf{M}_0^{2,2}&\mathbf{M}_0^{2,3}\\ \mathbf{0}&\mathbf{0}&\mathbf{M}_0^{3,3}\end{pmatrix},\qquad \mathbf{M}_1(\tau)=\begin{pmatrix}\mathbf{M}_1^{1,1}&\mathbf{M}_1^{1,2}&\mathbf{M}_1^{1,3}\\ \mathbf{0}&\mathbf{M}_1^{2,2}&\mathbf{M}_1^{2,3}\\ \mathbf{0}&\mathbf{0}&\mathbf{M}_1^{3,3}\end{pmatrix},$$

where

$$\mathbf{M}_0^{1,1}=\begin{pmatrix}0&\frac{3(z+1)}{z}&\frac{3(2z+1)}{z}\\ 0&-3&-6\\ 0&0&3\end{pmatrix},\qquad \mathbf{M}_0^{2,2}=\begin{pmatrix}-3&-\frac{6(z+1)}{z}&-\frac{3(z+2)}{z}\\ 0&3&3\\ 0&0&0\end{pmatrix},$$

$$\mathbf{M}_0^{1,2}=\begin{pmatrix}0&0&\frac{9(z+1)}{z}\\ 0&0&-9\\ 0&0&0\end{pmatrix},\qquad \mathbf{M}_0^{2,3}=\begin{pmatrix}0&-\frac{9}{z}&-\frac{9}{z}\\ 0&0&0\\ 0&0&0\end{pmatrix},$$

$$\mathbf{M}_0^{1,3}=\begin{pmatrix}0&0&\frac{9}{z}\\ 0&0&0\\ 0&0&0\end{pmatrix},\qquad \mathbf{M}_0^{3,3}=\begin{pmatrix}3&\frac{3(z+1)}{z}&-\frac{3(z-1)}{z}\\ 0&0&3\\ 0&0&-3\end{pmatrix},$$

$$\mathbf{M}_1^{1,1}=\begin{pmatrix}0&0&0\\ 0&3&0\\ 0&0&-3\end{pmatrix},\qquad \mathbf{M}_1^{2,2}=\begin{pmatrix}3&0&0\\ 0&-3&0\\ 0&0&0\end{pmatrix},$$

$$\mathbf{M}_1^{1,2}=\begin{pmatrix}-\frac{3(w+1)}{w}&0&0\\ 0&\frac{6(w+1)}{w}&\frac{9}{w}\\ 0&0&-\frac{3(w+1)}{w}\end{pmatrix},\quad \mathbf{M}_1^{2,3}=\begin{pmatrix}6&9&0\\ 0&-3&0\\ 0&0&-3\end{pmatrix},$$

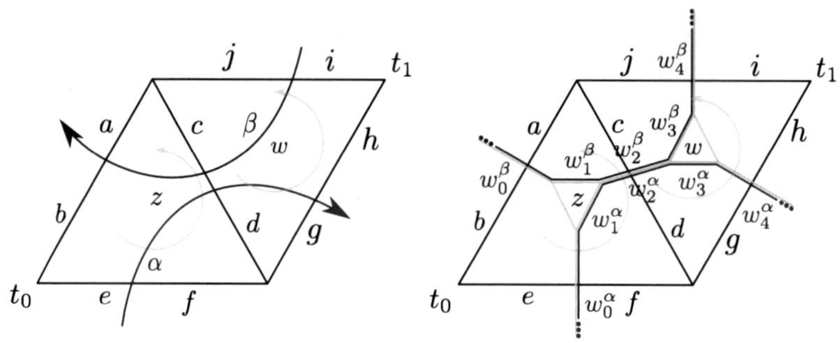

Figure 6.6: Example of type C.

$$\mathbf{M}_1^{1,3} = \begin{pmatrix} -\frac{3(2w+1)}{w} & -\frac{9(w+1)}{w} & -\frac{9}{w} \\ 0 & \frac{3(w+2)}{w} & \frac{9}{w} \\ 0 & 0 & \frac{3(w-1)}{w} \end{pmatrix}, \quad \mathbf{M}_1^{3,3} = \begin{pmatrix} -3 & 0 & 0 \\ 0 & 0 & 0 \\ 0 & 0 & 3 \end{pmatrix}.$$

Hence we apply Lemma 6.4.15 to check that $\mathbf{M}_{(0,1)}(\tau)$ is the zero matrix, which implies that the contribution along the tunnel of type C is zero. Since there is no point of intersection p in the tunnel $\epsilon(p, \alpha, \beta) = 0$ and the statement is proved for type C.

Cases A, B and D are treated verbatim, while for E, F, G and H one needs only to check that

$$\epsilon(p,\alpha,\beta)\frac{\mathbf{M}_{(0,1)}(\tau)}{9} = \begin{pmatrix} \begin{pmatrix} \frac{2}{3} & 0 & 0 \\ 0 & -\frac{1}{3} & 0 \\ 0 & 0 & -\frac{1}{3} \end{pmatrix} & \begin{pmatrix} 0 & 0 & 0 \\ 1 & 0 & 0 \\ 0 & 0 & 0 \end{pmatrix} & \begin{pmatrix} 0 & 0 & 0 \\ 0 & 0 & 0 \\ 1 & 0 & 0 \end{pmatrix} \\ \begin{pmatrix} 0 & 1 & 0 \\ 0 & 0 & 0 \\ 0 & 0 & 0 \end{pmatrix} & \begin{pmatrix} -\frac{1}{3} & 0 & 0 \\ 0 & \frac{2}{3} & 0 \\ 0 & 0 & -\frac{1}{3} \end{pmatrix} & \begin{pmatrix} 0 & 0 & 0 \\ 0 & 0 & 0 \\ 0 & 1 & 0 \end{pmatrix} \\ \begin{pmatrix} 0 & 0 & 1 \\ 0 & 0 & 0 \\ 0 & 0 & 0 \end{pmatrix} & \begin{pmatrix} 0 & 0 & 0 \\ 0 & 0 & 1 \\ 0 & 0 & 0 \end{pmatrix} & \begin{pmatrix} -\frac{1}{3} & 0 & 0 \\ 0 & -\frac{1}{3} & 0 \\ 0 & 0 & \frac{2}{3} \end{pmatrix} \end{pmatrix}.$$

It is easy to check that such matrix encodes precisely the contribution

$$\mathrm{tr}_{\alpha_p\beta_p} \circ \mu - \frac{1}{3}(\mathrm{tr}_\alpha \circ \mu)(\mathrm{tr}_\beta \circ \mu),$$

and the proof is complete. ■

Lemma 6.4.18. *The contribution along any tunnel equals*

$$\epsilon(p,\alpha,\beta)\left(\mathrm{tr}_{\alpha_p\beta_p} \circ \mu - \frac{1}{3}(\mathrm{tr}_\alpha \circ \mu)(\mathrm{tr}_\beta \circ \mu)\right),$$

where p is the point of intersection (if any) of α and β in the tunnel.

Proof. Let (w_α, w_β) be a tunnel τ through $(t_0, \ldots, t_q)$. Lemma 6.4.17 applies for $q = 1$. Hence assume $q > 1$. For all $0 < k < q$, the type of the transition triangle t_k depends on the relative orientation of α and β along the tunnel. In particular, if t_0 is a starting triangle of type $1, 1S$ (resp. $2, 4S$), then only transition triangles of type $5, 7$ (resp. $6, 6S$) are allowed.

Let $C_k(\tau)$ and $\mathbf{M}_k(\tau)$ be the contribution and the contribution matrix of the k–th gallery. Once can apply Lemma 6.4.15 to check that $\mathbf{M}_{(0,1)}(\tau) = \mathbf{M}_0(\tau)$ for all allowed type combinations of t_0 and t_1. It follows by an inductive argument that $\mathbf{M}_{(0,\ldots,k)}(\tau) = \mathbf{M}_0(\tau)$ for all $k < q$, where $\mathbf{M}_{(0,\ldots,k)}(\tau) \in M_{3\times3}(M_{3\times3}(\mathbb{R}))$ is the matrix contribution of the first k galleries, namely the matrix for which

$$\sum_{i=0}^{k} C_i(\tau) = \frac{2}{9}\sum_{l,m} \mathrm{tr}(\mathbf{E}_{lm} \cdot \mathbf{S}(w^\alpha_{2k+1})) \cdot \mathrm{tr}((\mathbf{M}_{(0,\ldots,k)}(\tau))_{lm} \cdot \mathbf{S}(w^\beta_{2k+1})).$$

In conclusion $\mathbf{M}_{(0,\ldots,q-1)}(\tau) = \mathbf{M}_0(\tau)$ and we are reduced to the case of length two tunnels, for which Lemma 6.4.17 applies. ■

6.4.8 Proof part 4: parallel curves and closed tunnels

When α and β are parallel, that is $[\alpha] = [\beta^n]$ for some $n \in \mathbb{Z}\backslash\{0\}$, a slightly different kind of tunnel appears.

We will say that (w_α, w_β) is a *closed tunnel through* $(t_0, \ldots, t_q)$ if t_j is a transition triangle with respect to the pairs $(w^\alpha_i, w^\beta_i)_{2j\leqslant i\leqslant 2j+2}$ for all $0 \leqslant j \leqslant q$, and $w^\gamma_0 = w^\gamma_{2k+2}$ for $\gamma \in \{\alpha, \beta\}$.

Lemma 6.4.19. *The contribution along a closed tunnel is* 0.

Proof. Let (w_α, w_β) be a closed tunnel τ through $(t_0, \ldots, t_q)$. We will show that (w_α, w_β) can be modified to an "open" tunnel with contribution 0. More precisely, we claim that the contribution along τ equals the contribution along the "open" tunnel $\widetilde{\tau} = (\widetilde{w_\alpha}, \widetilde{w_\beta})$ through $(t_0, \ldots, t_q, t_{q+1})$ defined by:

- $t_{q+1} := t_0$;
- $\widetilde{w_\alpha} := (w^\alpha_0 w^\alpha_1 w^\alpha_2 w^\alpha_3 \ldots w^\alpha_{2q+2} w^\alpha_1 w^\alpha_2)$;
- $\widetilde{w_\beta} := (\widetilde{w_0}\widetilde{w_1} w^\beta_2 w^\beta_3 \ldots w^\beta_{2q+2}\widetilde{w_2}\widetilde{w_3})$ where $\widetilde{w_0}, \widetilde{w_1}, \widetilde{w_2}, \widetilde{w_3}$ are as in Figure 6.7.

It is clear that $C_k(\tau) = C_k(\widetilde{\tau})$ for all $1 \leqslant k \leqslant q$, as we have only modified β locally at t_0. Nevertheless, the contribution of the 0–th gallery in (w_α, w_β) is

$$C_0(\tau) = \sum_{0\leqslant i,j\leqslant 2} \left(\sum_{q_1,q_2 \subset t_0} \epsilon(q_1, q_2) q^*_1 q^*_2 \left(\mathrm{tr}(w^\alpha_{1,i}) \cdot \mathrm{tr}(w^\beta_{2,j})\right)\right).$$

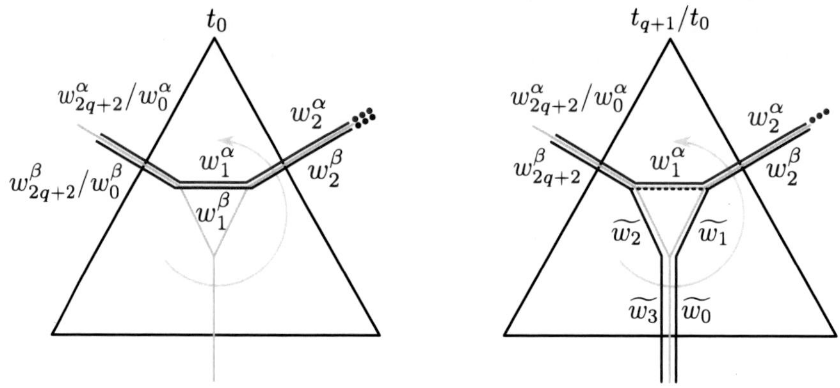

Figure 6.7: Local modification of β in a closed tunnel.

In such a gallery, we replace

$$\begin{aligned}\operatorname{tr}(w^{\beta}_{i,1}) &= \operatorname{tr}(S^{\beta}_0 \cdots \partial_i W^{\beta}_1 \cdots S^{\beta}_n)\\ &= \operatorname{tr}(S^{\beta}_0 \cdots \partial_i \left((W^{\beta}_1)^{-1}\widetilde{W}\,\widetilde{W}^{-1}(W^{\beta}_1)^{-1}\right) \cdots S^{\beta}_n)\\ &= \operatorname{tr}(\widetilde{w_{i,0}}) + \operatorname{tr}(\widetilde{w_{i,1}}) + \operatorname{tr}(\widetilde{w_{i,2}}) + \operatorname{tr}(\widetilde{w_{i,3}}),\end{aligned}$$

where $\widetilde{W}$ is the matrix associated to the $\underline{E}$–edge corresponding to $\widetilde{w_0}$ (and $\widetilde{w_3}$). It follows that $C_0(\tau) = C_0(\tilde{\tau}) + C_{q+1}(\tilde{\tau})$ and the claim is proved.

Finally, since $(\widetilde{w_\alpha}, \widetilde{w_\beta})$ is a tunnel with no intersection points, the conclusion is implied by Lemma 6.4.18. ■

6.4.9 Proof part 5: conclusion

Now it is time to put all the pieces together and prove Theorem 6.4.9. Recall that our goal is to show that

$$\begin{aligned}&\sum_{p\in\alpha\cap\beta} \epsilon(p,\alpha,\beta)\left(\operatorname{tr}_{\alpha_p\beta_p}\circ\mu - \frac{1}{3}(\operatorname{tr}_\alpha\circ\mu)(\operatorname{tr}_\beta\circ\mu)\right)\\ =&\sum_{j,k}\sum_{q_1,q_2\in\triangle\cup\underline{E}} \epsilon(q_1,q_2)q_1^*q_2^*\left(\operatorname{tr}(s^{\alpha}_{1,j})\cdot\operatorname{tr}(s^{\beta}_{2,k})\right).\end{aligned}$$

Firstly, assume α and β are not parallel. If (s^α_i, s^β_j) do not cross a common triangle, then we can discard such pair from the right hand sum as it contributes nothing. Otherwise, (s^α_i, s^β_j) is in the gallery of a unique tunnel. By Lemma 6.4.18, if α and β are not intersecting along such tunnel, then the sum of the contributions of all pairs in the tunnel is zero and we can discard them too. Hence we are left only with pairs which are contained in a tunnel where α and β do intersect. But since each intersection point is contained in a unique tunnel, we can apply Lemma 6.4.18 again to conclude the proof.

When α and β are parallel, the only difference from the previous argument is the possible presence of closed tunnels. However Lemma 6.4.19 ensures that there is no contribution from those tunnels so we may discard them and follow the above construction verbatim. This concludes the proof of Theorem 6.4.9.

Chapter 7

Examples

In this section we provide examples of the properties of $\mathcal{T}_3^+(S_{g,n})$ discussed throughout this paper. As usual, we denote by $S_{g,n}$ the oriented surface of genus g with n punctures, and we fix an ideal triangulation of $S_{g,n}$, whose set of oriented edges is $\underline{E}$ and whose set of triangles is $\triangle$.

7.1 The once-punctured torus

Let $S_{1,1}$ as depicted by the ideal triangulation T as in Figure 7.1. We remark that opposite edges of the rectangle are identified hence they display the same edge ratios.

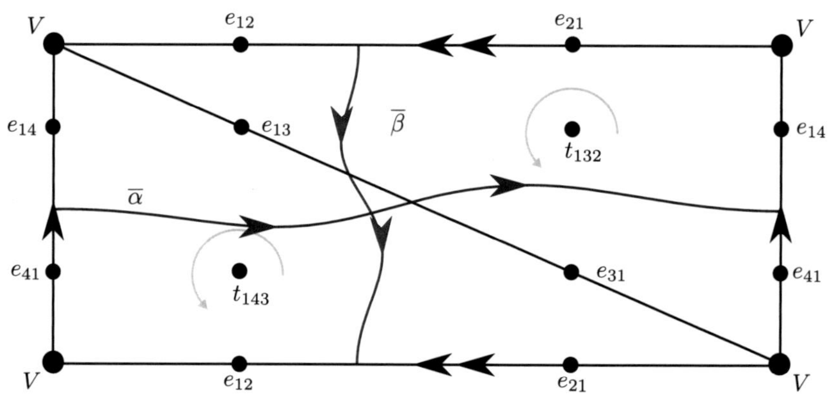

Figure 7.1: An ideal triangulation of $S_{1,1}$ with a single ideal vertex V.

Fock and Goncharov's Theorem 4.4.1 shows that the assignment of triple ratios and edge ratios indicated in Figure 7.1 uniquely determines an element of

$$\mathcal{T}_3^+(S_{1,1}) \cong \mathbb{R}_{>0}^{\triangle \cup \underline{E}} \cong \mathbb{R}_{>0}^8.$$

Moreover we can read off information about the corresponding holonomy group Γ with knowledge of the edge ratios and triple ratios alone using the method of §5.2.2. Let $\overline{\alpha}, \overline{\beta}$ be the generators of $\pi_1(S_{1,1})$ depicted in Figure 7.1. We denote by α and β a choice of holonomies of $\overline{\alpha}$ and $\overline{\beta}$ respectively. The method of §5.2.2 determines that, up to a common conjugation

$$\alpha = T(t_{143})E(e_{31}, e_{13})T(t_{132})E(e_{41}, e_{14})T(t_{143}),$$
$$\beta = E(e_{12}, e_{21})T(t_{132})E(e_{13}, e_{31})T(t_{143})^{-1},$$

where we recall that,

$$T(z) := \frac{1}{\sqrt[3]{z}}\begin{pmatrix} 0 & 0 & 1 \\ 0 & -1 & -1 \\ z & z+1 & 1 \end{pmatrix}, \qquad \text{and} \qquad E(x,y) := \sqrt[3]{\frac{x}{y}}\begin{pmatrix} 0 & 0 & y \\ 0 & -1 & 0 \\ \frac{1}{x} & 0 & 0 \end{pmatrix}.$$

A direct calculation of the generator $\alpha\beta\alpha^{-1}\beta^{-1}$ of the peripheral subgroup of $\mathrm{hol}(S_{1,1})$ yields the requisite information for further inquiry into the structure of $\mathcal{T}_3^+(S_{1,1})$. We have:

$$\alpha\beta\alpha^{-1}\beta^{-1} = \frac{1}{t_{132}t_{143}}\begin{bmatrix} (e_{12}e_{13}e_{21}e_{31}e_{41}e_{14})^{-1} & 0 & 0 \\ * & 1 & 0 \\ * & * & e_{12}e_{13}e_{21}e_{31}e_{41}e_{14}(t_{132}t_{143})^3 \end{bmatrix}.$$

This is the normal form we expect to arise following the construction in §5.2.2. Recall from §5.3, or by inspection of the above matrix, that the following monomials determine whether the end of $S_{1,1}$ is hyperbolic, special or cuspidal:

$$\mathcal{X}_1 := (t_{132}t_{143})^3 e_{12}e_{21}e_{14}e_{41}e_{13}e_{31},$$
$$\mathcal{Y}_1 := e_{12}e_{21}e_{14}e_{41}e_{13}e_{31}.$$

These monomials are our primary tools in identifying interesting subvarieties of $\mathcal{T}_3^+(S_{1,1})$. Let us begin by identifying which inequalities on coordinates give rise to structures which have minimal or maximal hyperbolic ends. This is a simple calculation following the classification in §5.3. The region in which coordinates give rise to structures on $S_{1,1}$ with a maximal hyperbolic end are:

$$\mathcal{X}_1 < \mathcal{Y}_1\mathcal{X}_1 < 1, \qquad \text{or} \qquad \mathcal{X}_1 > \mathcal{Y}_1\mathcal{X}_1 > 1.$$

Similarly, those structures with minimal hyperbolic ends are given by:

$$\mathcal{Y}_1 < 1 \quad \text{and} \quad \mathcal{X}_1 < 1, \quad \text{or} \quad \mathcal{X}_1\mathcal{Y}_1 < 1 \quad \text{and} \quad \mathcal{X}_1 > 1,$$
$$\text{or} \quad \mathcal{Y}_1 > 1 \quad \text{and} \quad \mathcal{X}_1 > 1, \quad \text{or} \quad \mathcal{X}_1\mathcal{Y}_1 > 1 \quad \text{and} \quad \mathcal{X}_1 < 1.$$

We see that generically (on structures with hyperbolic ends) there is a natural $2:1$ correspondence between structures with minimal hyperbolic ends and those with maximal hyperbolic ends, as noted in the discussion of the $\mathrm{Sym}(3)$–action on $\mathcal{T}_3^+(S_{1,1})$ in §4.5.

We will now identify the symplectic leaves of $\mathcal{T}_3^+(S_{1,1})$. Recall that, given a

Poisson bracket

$$\{-,-\} : C^\infty(M) \times C^\infty(M) \to C^\infty(M),$$

a *Casimir* is a smooth function $f \in C^\infty(M)$ such that $\{f, g\}$ is the zero function for all $g \in C^\infty(M)$. Following Goldman [17], or by a brief calculation, we find that $\mathcal{X}_1$ and $\mathcal{Y}_1$ are Casimirs for the Poisson bracket $\{-,-\}_\mathrm{T}$ defined on $\mathbb{R}_{>0}^{\triangle \cup \underline{E}} \cong \mathcal{T}_3^+(S_{1,1})$ in §6.4.

The condition of $\mathcal{X}_1$ and $\mathcal{Y}_1$ being constant simplifies to

$$t_{132}t_{143} = c_0, \quad \text{and} \quad e_{12}e_{21}e_{14}e_{41}e_{13}e_{31} = c_1, \tag{7.1.1}$$

for any positive real constants c_0 and c_1. In particular the symplectic leaves of $\mathcal{T}_3^+(S_{1,1})$ are open cells of 6 real dimensions defined by the conditions (7.1.1). Denote a generic level set of the equations (7.1.1) by $\mathcal{L}$. Suppose f is a Casimir in the Poisson algebra $(C^\infty(\mathcal{L}), \{-,-\}_\mathrm{T})$. A coordinate system on $\mathcal{L}$ is determined by the functions

$$\mathcal{P} := \{t_{143}^*, e_{12}^*, e_{21}^*, e_{14}^*, e_{41}^*, e_{13}^*\} \subset C^\infty(\mathcal{L}),$$

where q^* for $q \in \triangle \cup \underline{E}$ are coordinate functions on $\mathcal{T}_3^+(S_{1,1})$ defined in §6.4.3. Computing $\{f, q^*\}_\mathrm{T}$ for $q^* \in \mathcal{P}$ shows that $\partial f/\partial q^* = 0$ for all such q^*. In particular f is constant on $\mathcal{L}$. Therefore, $\{-,-\}_\mathrm{T}$ is non-degenerate on $\mathcal{L}$ in the sense that the function f is uniformly the zero function if and only if f is constant. It follows that $\mathcal{L}$ is a symplectic leaf.

Increasing once more in specificity, recall from §6.1 that the finite-area structures on $S_{1,1}$ arise when $\mathcal{X}_1 = \mathcal{Y}_1 = 1$. Therefore the finite-area conditions for $S_{1,1}$ reduce to the particularly simple condition that $c_0 = c_1 = 1$ in equations (7.1.1).

We now determine $\mathrm{Teich}(S_{1,1}) \subset \mathcal{T}_3^+(S_{1,1})$. Recall that the conditions for Fock-Goncharov coordinates to represent a finite-area hyperbolic structure described in §6.2 are:

$$\phi(x)(e) = \phi(x)(e^{-1}), \quad \forall\, e \in \underline{E}, \quad \text{and} \quad \phi(x)(t) = 1, \quad \forall\, t \in \triangle,$$
$$\text{and} \quad \prod_{k=1}^{m_i} \phi(e_{ik}) = 1, \qquad \forall\, i = 1, \ldots, n.$$

In the present example this simplifies to

$$t_{143} = t_{132} = 1, \qquad e_{ij} = e_{ji}, \quad \forall i, j, \quad \text{and} \qquad e_{21}e_{41}e_{31} = 1,$$

So $\mathrm{Teich}(S_{1,1}) \cong \mathbb{R}^2$ as we should expect from the classical theory.

7.2 The thrice-punctured sphere

We use the same setup as above with $S_{0,3}$. Just as in the previous example, Figure 7.2 denotes an ideal triangulation T of $S_{0,3}$. Once again, the edge ratios and triple ratios uniquely determine an element of

$$\mathcal{T}_3^+(S_{0,3}) \cong \mathbb{R}_{>0}^{\triangle \cup E} \cong \mathbb{R}_{>0}^8.$$

The surface $S_{0,3}$ has fundamental group isomorphic to the free group on two genera-

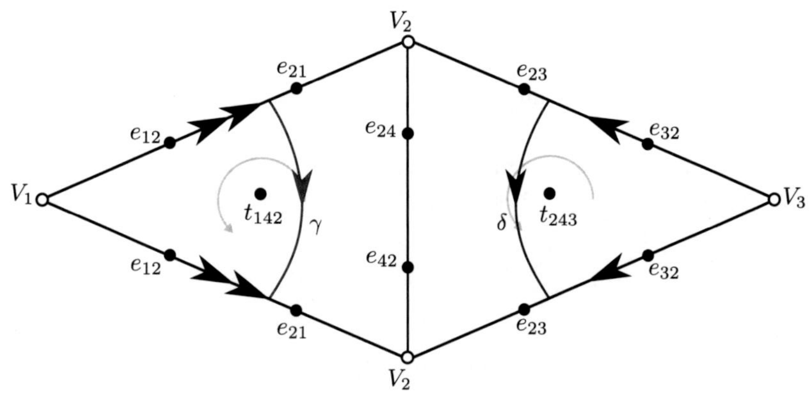

Figure 7.2: An ideal triangulation of $S_{0,3}$ with two ideal triangles, three ideal edges and three ideal vertices $\{V_i\}_{i=1,2,3}$.

tors. The holonomy group Γ is generated by the holonomies of deck transformations corresponding to lifts of the face pairings which identify the two edges on the left and the two edges on the right of Figure 7.2. We denote these by γ and $\delta \in \Gamma$, respectively. Their $\mathrm{SL}(3,\mathbb{R})$–representatives are given below, up to a common conjugation, using the method of §5.2.2:

$$\gamma = E(e_{12}, e_{21})T(t_{142}) = \sqrt[3]{\frac{e_{12}}{t_{142}e_{21}}} \begin{bmatrix} t_{142}e_{21} & e_{21}(t_{142}+1) & e_{21} \\ 0 & 1 & 1 \\ 0 & 0 & \frac{1}{e_{12}} \end{bmatrix},$$

$$\delta = T(t_{142})E(e_{42}, e_{24})T(t_{243})E(e_{23}, e_{32})T(t_{243})E(e_{24}, e_{42})T(t_{142})^{-1}.$$

We have chosen a conjugacy class of the holonomy group so that γ is in a convenient form. However, as there are three distinct peripheral subgroups of $\pi_1(S_{0,3})$, the other generators of these subgroups, δ and $\gamma \cdot \delta^{-1}$ are not written as succinctly. On the other hand, their eigenvalues are still easily calculable. For more complicated examples it is more convenient to make use of Theorem 5.2.4 to infer that the following monomials govern the boundary geometry of $S_{0,3}$:

$$\begin{aligned} \gamma : &\quad \mathcal{X}_1 := t_{142}e_{21}, &\quad \text{and} &\quad \mathcal{Y}_1 := e_{12}, \\ \gamma \cdot \delta^{-1} : &\quad \mathcal{X}_2 := t_{243}^2 t_{142}^2 e_{32}e_{42}e_{12}e_{24}, &\quad \text{and} &\quad \mathcal{Y}_2 := e_{23}e_{24}e_{21}e_{42}, \\ \delta : &\quad \mathcal{X}_3 := t_{243}e_{23}, &\quad \text{and} &\quad \mathcal{Y}_3 := e_{32}. \end{aligned}$$

Therefore there are at least six independent Casimirs described by the equations $\mathcal{X}_i = c_i$ and $\mathcal{Y}_i = d_i$ for positive real constants $\{c_i, d_i\}_{i=1,2,3}$. The equations defining the level sets of these Casimirs simplify to:

$$\begin{aligned} e_{12} =& d_1, \quad && \text{and} \quad && t_{142}e_{21} = c_1, \\ e_{23}e_{24}e_{21}e_{42} =& d_2, \quad && \text{and} \quad && e_{23}e_{21} = \left(\frac{d_3d_2d_1(c_1c_3)^2}{c_2}\right)^{\frac{1}{3}}, \\ e_{32} =& d_3, \quad && \text{and} \quad && t_{243}e_{23} = c_3. \end{aligned}$$

These equations are clearly independent so the maximal symplectic leaves are at most 2–dimensional. Symplectic manifolds have even dimension so the symplectic leaves have dimension either 0 or 2 in this case. The former occurs if and only if $\{-,-\}_{\mathrm{T}}$ is uniformly zero. A direct calculation (cf. §6.4.3) shows that

$$\{t^*_{142}, e^*_{42}\}_{\mathrm{T}} = 2t^*_{142}e^*_{42} \neq 0.$$

The subvarieties defined by level sets of the six Casimirs above must therefore determine symplectic leaves of the Poisson structure on $\mathbb{R}^{\triangle\cup\underline{E}}_{>0}$.

We have now seen that the maximal symplectic leaves of $\mathcal{T}_3^+(S_{0,3})$ are 2–dimensional whereas the symplectic leaves of $\mathcal{T}_3^+(S_{1,1})$ are 6–dimensional. This allows us to distinguish between $\mathcal{T}_3^+(S_{1,1})$ and $\mathcal{T}_3^+(S_{0,3})$ using their intrinsic Poisson geometry despite the fact that they are homeomorphic. These examples illustrate Theorem 6.4.4.

The structures on $S_{0,3}$ with finite area are described by $c_i = d_i = 1$ for $i = 1, 2, 3$ so $\mathcal{T}_3^f(S_{0,3})$ is defined by the conditions

$$e_{12} = e_{32} = 1, \quad e_{42}e_{24} = e_{23}e_{21} = 1, \quad \text{and} \quad e_{21}t_{142} = e_{23}t_{243} = 1.$$

Therefore $\mathcal{T}_3^f(S_{0,3}) \cong \mathbb{R}^3$. Imposing the finite-area conditions and the criteria presented in §6.2 shows that the classical Teichmüller space is determined by:

$$e_{ij} = 1, \quad \forall\, i, j, \qquad \text{and} \qquad t_{142} = t_{243} = 1.$$

It follows that $\mathrm{Teich}(S_{0,3})$ is a single point, as expected from the classical theory.

7.3 The twice-punctured torus

We consider the ideally triangulated twice-punctured torus $S_{1,2}$ depicted in Figure 7.3.

The four Casimirs arising from the peripheral elements of $\pi_1(S_{1,2})$ are determined by the monomials $\mathcal{X}_i$ and $\mathcal{Y}_i$ for $i = 1, 2$ as defined in §5.3. They are given explicitly as follows:

$$\begin{aligned} \mathcal{X}_1 :=& t_{162}t^2_{253}t_{265}t^2_{354}e_{21}e_{61}e_{53}e_{23}e_{25}e_{35}, \quad && \text{and} \quad && \mathcal{Y}_1 := e_{12}e_{16}e_{35}e_{32}e_{52}e_{53}, \\ \mathcal{X}_2 :=& t^2_{162}t^2_{265}t_{253}t_{354}e_{12}e_{62}e_{52}e_{32}e_{26}e_{16}, \quad && \text{and} \quad && \mathcal{Y}_2 := e_{21}e_{26}e_{25}e_{23}e_{62}e_{61}. \end{aligned}$$

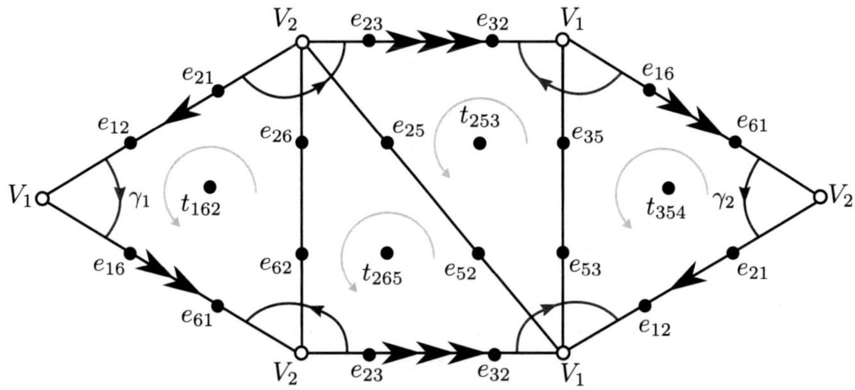

Figure 7.3: An ideal triangulation of $S_{1,2}$ with two ends V_i, for $i = 1, 2$.

The boundary geometry of $S_{1,2}$ is determined by the monomials $\mathcal{X}_i = c_i$ and $\mathcal{Y}_i = d_i$ for positive real constants c_i and d_i for $i = 1, 2$.

Setting $c_i = d_i = 1$ for $i = 1, 2$ we find that $\mathcal{T}_3^f(S_{1,2})$ is defined by

$$e_{12}e_{16}e_{35}e_{32}e_{52}e_{53} = 1, \quad \text{and} \quad t_{162}t_{253}^2t_{265}t_{354}^2e_{53}e_{35} = e_{62}e_{26},$$
$$e_{21}e_{26}e_{25}e_{23}e_{62}e_{61} = 1, \quad \text{and} \quad t_{162}^2t_{265}^2t_{253}t_{354}e_{26}e_{62} = e_{53}e_{35}.$$

It follows that $\mathcal{T}_3^f(S_{1,2}) \cong \mathbb{R}^{12}$.

Imposing both the conditions for finite-area and hyperbolicity as seen in §6.2 we identify Teich$(S_{1,2})$ with the subvariety defined by:

$$e_{ij} = e_{ji}, \quad \forall\, i,j, \quad \text{and} \quad t_{ijk} = 1, \quad \forall\, i,j,k,$$
$$e_{53}^2e_{21}e_{61}e_{23}e_{25} = 1, \quad \text{and} \quad e_{62} = e_{53}.$$

Therefore Teich$(S_{1,2}) \cong \mathbb{R}^4$ in accordance with the classical theory.

7.4 The closed genus two surface

Now we will consider the closed genus two surface $S_{2,0}$ and use the method of §6.3 to construct $\mathcal{T}_3(S_{2,0})$ from $\mathcal{T}_3^+(S_{1,2})$, by gluing boundary components.

In order to glue the boundary components of $S_{1,2}$ in the manner of Goldman [21], we must ensure that both ends of $S_{1,2}$ are minimal. Using the notation from the previous example, the set of structures whose i–th end is minimal ($i = 1, 2$) is the union of following four disjoint sets:

$$\{\mathcal{Y}_i > 1, \mathcal{X}_i > 1\} \cup \{\mathcal{Y}_i < 1, \mathcal{X}_i < 1\} \cup \{\mathcal{X}_i\mathcal{Y}_i < 1, \mathcal{X}_i > 1\} \cup \{\mathcal{X}_i\mathcal{Y}_i > 1, \mathcal{X}_i < 1\}. \tag{7.4.1}$$

Note that $\mathcal{T}_3(S_{2,0})$ does not contain information regarding framing so we will 'forget' the framing information contained in $\mathcal{T}_3^+(S_{1,2})$. This is tantamount to restricting our attention to exactly one of the regions defined by (7.4.1) for each boundary component. Hence we assume without loss of generality to have a structure in the first set for both $i = 1$ and $i = 2$, namely $\{\mathcal{Y}_i > 1, \mathcal{X}_i > 1\}$.

The second requirement in performing convex gluing is that the peripheral transformations at the ends of $S_{1,2}$ be conjugate. Therefore, following the classification of peripheral elements and their eigenvalues in §5.3, we require that

$$\mathcal{X}_1\mathcal{Y}_1^2 = \mathcal{X}_2\mathcal{Y}_2^2, \qquad \text{and} \qquad \mathcal{X}_1^2\mathcal{Y}_1 = \mathcal{X}_2^2\mathcal{Y}_2.$$

Given that all variables must be positive and real, we have one pair of solutions of the form $\mathcal{X}_1 = \mathcal{X}_2$ and $\mathcal{Y}_1 = \mathcal{Y}_2$. We obtain the following conditions which parameterise convex projective structures on $S_{1,2}$ whose ends are minimal with totally hyperbolic holonomy:

$$e_{12}e_{16}e_{35}e_{32}e_{52}e_{53} = e_{21}e_{26}e_{25}e_{23}e_{62}e_{61} > 1, \tag{7.4.2}$$

$$(e_{53}e_{35})^2 t_{354}t_{253} = (e_{26}e_{62})^2 t_{162}t_{265} > 1. \tag{7.4.3}$$

It remains to show that these equations define a topological cell of real dimension 16, thus verifying Goldman's [21] result. Following the method used in Theorem 6.1.4, it is sufficient to note that the equations (7.4.2)–(7.4.3) are independent. Recall that as long as the peripheral elements of $S_{1,2}$ are hyperbolic, there is a 2–real parameter space of potential gluings for each pair of 'gluable' ends. For an explanation of this see §6.3 or the original proof in Goldman [21]. Therefore $\mathcal{T}_3(S_{2,0})$ is a trivial 2–dimensional vector bundle over the subvariety of $\mathcal{T}_3^+(S_{1,2})$ defined by the conditions (7.4.2)–(7.4.3). Ultimately this shows that $\mathcal{T}_3(S_{2,0}) \cong \mathbb{R}^{16}$, reproducing Goldman's result.

Once again we can use this to calculate a subvariety of $\mathcal{T}_3(S_{2,0})$ which is homeomorphic to Teich$(S_{2,0})$. First note that any hyperbolic structure on $S_{2,0}$ must restrict to a hyperbolic structure on $S_{1,2}$. Therefore we impose the hyperbolicity conditions on $\mathcal{T}_3^+(S_{1,2})$ defined in §6.2 to obtain:

$$t_{ijk} = 1, \quad \forall\, i,j,k, \qquad \text{and} \qquad e_{ij} = e_{ji} \quad \forall\, i,j.$$

Further imposing these conditions on the 'gluable' equations we are left with:

$$e_{35} = e_{62}, \quad t_{ijk} = 1, \quad \forall\, i,j,k, \quad \text{and} \quad e_{ij} = e_{ji}, \quad \forall\, i,j. \tag{7.4.4}$$

To recover Teich$(S_{2,0})$ note that instead of a 2–parameter family of convex gluings we may only glue the ends of $S_{1,2}$ by hyperbolic isometries if we wish to give rise to a hyperbolic structure on $S_{2,0}$. Therefore, Teich$(S_{2,0})$ may be realised as a trivial real 1-dimensional vector bundle over the subvariety defined by the conditions (7.4.4). In particular Teich$(S_{2,0}) \cong \mathbb{R}^6$, which accords with the classical theory.

References

[1] Michèle Audin. Lectures on gauge theory and integrable systems. In *Gauge theory and symplectic geometry (Montreal, PQ, 1995)*, volume 488 of *NATO Adv. Sci. Inst. Ser. C Math. Phys. Sci.*, pages 1–48. Kluwer Acad. Publ., Dordrecht, 1997. 8

[2] Jean-Paul Benzécri. Sur les variétés localement affines et localement projectives. *Bull. Soc. Math. France*, 88:229–332, 1960. 2, 32, 34

[3] Francis Bonahon and Guillaume Dreyer. Parameterizing Hitchin components. *Duke Math. J.*, 163(15):2935–2975, 2014. 4, 83

[4] Martin R. Bridson and André Haefliger. *Metric spaces of non-positive curvature*, volume 319 of *Grundlehren der Mathematischen Wissenschaften [Fundamental Principles of Mathematical Sciences]*. Springer-Verlag, Berlin, 1999. 29

[5] Herbert Busemann and Paul J. Kelly. *Projective geometry and projective metrics.* Academic Press Inc., New York, N. Y., 1953. 16, 28, 30

[6] James W. Cannon, William J. Floyd, Richard Kenyon, and Walter R. Parry. Hyperbolic geometry. In *Flavors of geometry*, volume 31 of *Math. Sci. Res. Inst. Publ.*, pages 59–115. Cambridge Univ. Press, Cambridge, 1997. 29

[7] Suhyoung Choi and William M. Goldman. Convex real projective structures on closed surfaces are closed. *Proc. Amer. Math. Soc.*, 118(2):657–661, 1993. 2

[8] Suhyoung Choi and William M. Goldman. The classification of real projective structures on compact surfaces. *Bull. Amer. Math. Soc. (N.S.)*, 34(2):161–171, 1997. 85

[9] Daryl Cooper and Kelly Delp. The marked length spectrum of a projective manifold or orbifold. *Proc. Amer. Math. Soc.*, 138(9):3361–3376, 2010. 59

[10] Daryl Cooper, Darren Long, and Stephan Tillmann. On convex projective manifolds and cusps. *Adv. Math.*, 277:181–251, 2015. 28, 47, 48

[11] Daryl Cooper, Darren Long, and Stephan Tillmann. Deforming convex projective manifolds. *Geom. Topol.*, 22(3):1349–1404, 2018. 20, 24, 32

[12] Pierre de la Harpe. On Hilbert's metric for simplices. In *Geometric group theory, Vol. 1 (Sussex, 1991)*, volume 181 of *London Math. Soc. Lecture Note Ser.*, pages 97–119. Cambridge Univ. Press, Cambridge, 1993. 28, 29, 30

[13] William Floyd and Allen Hatcher. Incompressible surfaces in punctured-torus bundles. *Topology Appl.*, 13(3):263–282, 1982. 50

[14] Vladimir Fock and Alexander Goncharov. Moduli spaces of local systems and higher Teichmüller theory. *Publ. Math. Inst. Hautes Études Sci.*, (103):1–211, 2006. 2, 4

[15] Vladimir V. Fock and Alexander B. Goncharov. Moduli spaces of convex projective structures on surfaces. *Adv. Math.*, 208(1):249–273, 2007. 2, 5, 6, 7, 8, 35, 37, 42, 46, 53, 61

[16] Thomas Foertsch and Anders Karlsson. Hilbert metrics and Minkowski norms. *J. Geom.*, 83(1-2):22–31, 2005. 30

[17] William M. Goldman. The symplectic nature of fundamental groups of surfaces. *Adv. in Math.*, 54(2):200–225, 1984. 8, 28, 86, 87, 110

[18] William M. Goldman. Invariant functions on Lie groups and Hamiltonian flows of surface group representations. *Invent. Math.*, 85(2):263–302, 1986. 88

[19] William M. Goldman. Geometric structures on manifolds and varieties of representations. In *Geometry of group representations (Boulder, CO, 1987)*, volume 74 of *Contemp. Math.*, pages 169–198. Amer. Math. Soc., Providence, RI, 1988. 20, 24

[20] William M. Goldman. *Projective geometry on manifolds.* 1988. Notes available from the author. 59

[21] William M. Goldman. Convex real projective structures on compact surfaces. *J. Differential Geom.*, 31(3):791–845, 1990. 2, 4, 83, 113, 114

[22] François Guéritaud, Olivier Guichard, Fanny Kassel, and Anna Wienhard. Compactification of certain Clifford-Klein forms of reductive homogeneous spaces. *Michigan Math. J.*, 66(1):49–84, 2017. 4

[23] Olivier Guichard and Anna Wienhard. Convex foliated projective structures and the Hitchin component for $\mathrm{PSL}_4(\mathbf{R})$. *Duke Math. J.*, 144(3):381–445, 2008. 4

[24] Krishnamurthi Guruprasad, Johannes Huebschmann, Lisa Jeffrey, and Alan Weinstein. Group systems, groupoids, and moduli spaces of parabolic bundles. *Duke Math. J.*, 89(2):377–412, 1997. 8, 86

[25] Robert C. Haraway, III and Stephan Tillmann. Tessellating the moduli space of strictly convex projective structures on the once-punctured torus. *Exp. Math.*, 28(3):369–384, 2019. 3, 62, 70

[26] Robin Hartshorne. *Algebraic geometry.* Springer-Verlag, New York-Heidelberg, 1977. Graduate Texts in Mathematics, No. 52. 20, 87

[27] Allen Hatcher. On triangulations of surfaces. *Topology Appl.*, 40(2):189–194, 1991. 50

[28] Morris W. Hirsch. *Differential topology.* Springer-Verlag, New York-Heidelberg, 1976. Graduate Texts in Mathematics, No. 33. 23

[29] John Hamal Hubbard. *Teichmüller theory and applications to geometry, topology, and dynamics. Vol. 1.* Matrix Editions, Ithaca, NY, 2006. Teichmüller theory, With contributions by Adrien Douady, William Dunbar, Roland Roeder, Sylvain Bonnot, David Brown, Allen Hatcher, Chris Hruska and Sudeb Mitra, With forewords by William Thurston and Clifford Earle. 51

[30] Michael Kapovich, Bernhard Leeb, and Joan Porti. Dynamics on flag manifolds: domains of proper discontinuity and cocompactness. *Geom. Topol.*, 22(1):157–234, 2018. 4

[31] David C. Kay. The ptolemaic inequality in Hilbert geometries. *Pacific J. Math.*, 21:293–301, 1967. 31, 81

[32] Nicolaas H. Kuiper. Sur les surfaces localement affines. In *Géométrie différentielle. Colloques Internationaux du Centre National de la Recherche Scientifique, Strasbourg, 1953*, pages 79–87. Centre National de la Recherche Scientifique, Paris, 1953. 2

[33] Nicolaas H. Kuiper. On convex locally-projective spaces. In *Convegno Internazionale di Geometria Differenziale, Italia, 1953*, pages 200–213. Edizioni Cremonese, Roma, 1954. 2

[34] François Labourie. Cross ratios, surface groups, $\mathrm{PSL}(n, \mathbf{R})$ and diffeomorphisms of the circle. *Publ. Math. Inst. Hautes Études Sci.*, (106):139–213, 2007. 4

[35] Sean Lawton. Poisson geometry of $\mathrm{SL}(3,\mathbb{C})$-character varieties relative to a surface with boundary. *Trans. Amer. Math. Soc.*, 361(5):2397–2429, 2009. 87, 88

[36] Walter Lawrence Lok. *Deformations of locally homogeneous spaces and Kleinian groups (Teichmüller, hyperbolic)*. ProQuest LLC, Ann Arbor, MI, 1984. Thesis (Ph.D.)–Columbia University. 24

[37] Ludovic Marquis. Espace des modules marqués des surfaces projectives convexes de volume fini. *Geom. Topol.*, 14(4):2103–2149, 2010. 2, 4, 32, 79

[38] Ludovic Marquis. Surface projective convexe de volume fini. *Ann. Inst. Fourier (Grenoble)*, 62(1):325–392, 2012. 24, 76, 77

[39] Vladimir S. Matveev and Marc Troyanov. Isometries of two dimensional Hilbert geometries. *Enseign. Math.*, 61(3-4):453–460, 2015. 28, 30

[40] Lee Mosher. Tiling the projective foliation space of a punctured surface. *Trans. Amer. Math. Soc.*, 306(1):1–70, 1988. 50

[41] Xin Nie. *Structures quasi-Poisson sur l'espace des modules de connexions plates, entropies d'ensembles de Tits simpliciaux*. ProQuest LLC, Ann Arbor, MI, 2013. Thesis (Ph.D.)–École Doctorale Paris Centre. 8

[42] Anne Parreau. Invariant subspaces for some surface groups acting on A2-Euclidean buildings. *ArXiv e-prints*, April 2015. 65

[43] Claudio Procesi. The invariant theory of $n\times n$ matrices. *Advances in Math.*, 19(3):306–381, 1976. 88

[44] Édith Socié-Méthou. *Comportements asymptotiques et rigidites en geometries de Hilbert*. 2000. Thesis (Ph.D.)–Université de Strasbourg. 31

[45] Elias M. Stein and Rami Shakarchi. *Real analysis*, volume 3 of *Princeton Lectures in Analysis*. Princeton University Press, Princeton, NJ, 2005. Measure theory, integration, and Hilbert spaces. 31

[46] Dennis Sullivan and William and Thurston. Manifolds with canonical coordinate charts: some examples. *Enseign. Math. (2)*, 29(1-2):15–25, 1983. 21

[47] Oswald Teichmüller. Extremale quasikonforme Abbildungen und quadratische Differentiale. *Abh. Preuss. Akad. Wiss. Math.-Nat. Kl.*, 1939(22):197, 1940. 81, 83

[48] William P. Thurston. *Three-dimensional geometry and topology. Vol. 1*, volume 35 of *Princeton Mathematical Series*. Princeton University Press, Princeton, NJ, 1997. Edited by Silvio Levy. 19, 29, 51

[49] Constantin Vernicos. Introduction aux géométries de Hilbert. In *Actes de Séminaire de Théorie Spectrale et Géométrie. Vol. 23. Année 2004–2005*, volume 23 of *Sémin. Théor. Spectr. Géom.*, pages 145–168. Univ. Grenoble I, Saint-Martin-d'Hères, 2005. 30

[50] Èrnest B. Vinberg. The theory of homogeneous convex cones. *Trudy Moskov. Mat. Obšč.*, 12:303–358, 1963. 59

[51] Èrnest B. Vinberg and Victor G. Kac. Quasi-homogeneous cones. *Mat. Zametki*, 1:347–354, 1967. 2

[52] Hassler Whitney. Differentiable manifolds. *Ann. of Math. (2)*, 37(3):645–680, 1936. 3

Index